I0050918

Matthew N. O. Sadiku
Emerging Technologies in Africa

Integrated Global STEM

———

Edited by
Robert Krueger, Wole Soboyejo and Anita Mattson

Volume 5

Matthew N. O. Sadiku

Emerging Technologies in Africa

—

DE GRUYTER   The WPI Press

Author
Matthew N. O. Sadiku, Ph.D., P.E.
Regents Professor Emeritus and IEEE Life Fellow
Prairie View A&M University
Prairie View, TX 77446
U.S.A.
Email: sadiku@ieee.org
Web: www.matthew-sadiku.com

ISBN 978-3-11-914583-1
e-ISBN (PDF) 978-3-11-221198-4
e-ISBN (EPUB) 978-3-11-221255-4

Library of Congress Control Number: 2025940832

Bibliographic information published by the Deutsche Nationalbibliothek
The Deutsche Nationalbibliothek lists this publication in the Deutsche Nationalbibliografie;
detailed bibliographic data are available on the internet at http://dnb.dnb.de.

© 2025 Walter de Gruyter GmbH, Berlin/Boston, Genthiner Straße 13, 10785 Berlin
Cover image: burcu demir/iStock/Getty Images Plus
Typesetting: Integra Software Services Pvt. Ltd.

www.degruyterbrill.com
Questions about General Product Safety Regulation:
productsafety@degruyterbrill.com

DEDICATED TO MY LATE WIFE
CHRISTIANAH YEMISI

Preface

Africa is the second largest continent in the world after Asia and the second largest continent in population. It's a continent that has 54 countries with an area of 30,370,000 square km, subdivided into five major regions. More than 1.4 billion people (as of 2021) live on the African continent, which implies that about 15% of the world's total population live in Africa. Many scientists consider Africa to be the origin of mankind with cultural diversity. It is home to incredible wildlife and all kinds of unique, healthy, and delicious foods. It is a continent rich in resources, especially food-related products. As multiple nations and cultures make up Africa, the cuisine is each nation changes considerably. There are 54 countries in Africa. Africa is a continent that is rapidly embracing cutting-edge technology to leapfrog into the future.

Every modern society relies heavily on technology. Technology is everywhere. It surrounds every aspect of twenty-first-century life. It is in the cell phones we use, the cars we drive, and even the food we eat. Technology connects and brings us together to share ideas across borders and cultures. Emerging technologies can help Africa improve industrialization, thereby increasing economic growth and development. They promise to reduce the costs of undertaking research, offer endless opportunities for innovations and present opportunities to catch up with leading nations. These technologies can enable economic transformation in Africa and help create more jobs for its people. The deployment of cutting-edge technologies such as artificial intelligence and big data has the potential to revolutionize various sectors, including healthcare, education, agriculture, and finance, and drive sustainable development and economic growth across Africa.

This book explores the use of emerging technologies in Africa. These technologies include artificial intelligence, robotics, 3D printing, cloud computing, nanotechnology, Internet of things, blockchain, and drones. The book is organized into ten chapters that summarize these technologies and their adoption in African nations.

Chapter 1: Introduction: This chapter highlights the opportunities and challenges of implementing emerging technologies in Africa. "Emerging technology" is a general term used to describe new technology. They refer to technologies that are still being developed and play a key role in industrial modernization. They include artificial intelligence, robotics, 3D printing, cloud computing, nanotechnology, Internet of things, blockchain, and drones. The chapter unpacks what some of the economies in Africa have achieved in terms of emerging technologies. African countries that take advantage of the emerging technologies and limit the risks inherent in them in order to achieve greater peace and prosperity.

Chapter 2: Artificial Intelligence in Africa: This chapter examines the adoption of artificial intelligence (AI) in African nations. Artificial intelligence, sometimes called machine intelligence, refers to intelligence demonstrated by machines. Typically, AI systems demonstrate at least some of the following human behaviors: planning,

https://doi.org/10.1515/9783112211984-202

learning, reasoning, problem solving, knowledge representation, perception, speech recognition, decision-making, language translation, motion, manipulation, intelligence, and creativity. AI has the potential to fundamentally change the way businesses operate, drive innovation, and improve the lives of millions of people across Africa. Africa is a place where AI is used and developed: in social networks, in businesses, in healthcare, in agriculture, and in education. Africa is poised to lead and significantly influence the AI landscape.

Chapter 3: Robotics in Africa: This chapter examines the adoption of robotics in African nations. Robotics constitutes one of the most exciting fields of technology today. It is the discipline of designing and constructing intelligent machines, called robots. Robots are becoming increasingly prevalent in almost every industry, from healthcare to manufacturing. In recent years, several organizations in Africa have launched initiatives to advance participation in robotics. In some parts of Africa, robots are used in agriculture, manufacturing, and education. Africa is getting ready for the impending age of robots.

Chapter 4: Drones in Africa: In this chapter, we explore the adoption of drones in African nations. Drones are autonomous or remotely controlled multipurpose aerial vehicles driven by aerodynamic forces. They may be regarded as autonomous robots that fly in the sky. The usage of drone technology is gaining attention in Africa. Africans are employing drones for commercial, humanitarian, military purposes, and emergency medical situations. Drone technology is providing the delivery solutions that can enable African nations distribute essential medical supplies to rural areas. The drone industry in Africa is taking off, and it is evolving into a massive enterprise.

Chapter 5: Big Data in Africa: This chapter examines the impact and prospects of big data in developing African nations. We live in an era of data. The world's most valuable resource is no longer oil, but data. "Big data" is an umbrella term for large volumes of structured and unstructured data that we encounter daily. It refers to vast volumes of data that is too large for ordinary computing devices to process. Big data applications have the potential to catalyze the solution of local development challenges in African nations, including applications in key sectors of the economy such as agriculture, healthcare, energy, business, and resource management. Big data can help African countries improve their governance, reduce corruption, and increase transparency in public services.

Chapter 6: Cloud Computing in Africa: In this chapter, we consider the use of cloud computing in developing African nations. Cloud computing is a computing paradigm for delivering computing services (such as servers, storage, databases, networking, software, analytics, and more) over the "the cloud" or Internet with pay-as-you-go pricing. It is one of the globally recognized emerging technologies in the new millennium that are most likely to change people's lives. Rather than building, owning, and maintaining their own IT infrastructure, businesses can use cloud to access technology resources such as computing capacity, storage, and databases on a pay-as-you-go basis. The African continent is set for significant cloud adoption. Mobility is a major driver of cloud implementations in Africa. With cloud infrastructure, African businesses can compete on a global stage, offering services and products beyond local markets.

Chapter 7: Internet of Things in Africa: This chapter addresses the adoption of the Internet of things (IoT) in African nations. Today, the Internet has become an indispensable part of life. When it comes to the Internet, the IoT has taken center stage. The Internet of things refers to the billions of physical devices connected to the Internet that allows exchanging data around the world. It is one of the disruption technologies and is growing rapidly. It is an integral part of Future Internet. Although Africa is still behind the rest of the world in terms of Internet penetration, the gap is quickly closing. The future of IoT in Africa is bright and promising, with significant potential to transform various sectors such healthcare and agriculture.

Chapter 8: Blockchain in Africa: This chapter covers the use or adoption of blockchain technology in Africa and its potential benefits for the region. Blockchain is a relatively new and exciting way of recording transactions in the digital age. Originally developed as the accounting method for the virtual cryptocurrency Bitcoin, blockchains are appearing in a variety of commercial applications today. It has been heralded as a "game changer" for the development of African economies. At the moment, Africa is one of the fastest-growing cryptocurrency markets in the world. The block chain technology continues to impact Africa to a degree that is unprecedented anywhere else in the world. With continued investment, collaboration, and education, Africa has the opportunity to become a global leader in blockchain technology.

Chapter 9: 3D Printing in Africa: This chapter explores the state of art of 3D printing technology in Africa. 3D printing is a type of industrial robotics. It is also known as additive manufacturing or rapid prototyping. It is the process for making a physical object from a three-dimensional computer-aided design (CAD) file via a layering approach. Adopting 3D printing can strengthen the manufacturing industry in Africa. African nations need to invest in infrastructure, educated personnel, industry professionals, R&D, and innovation institutions to enhance the growth rate of 3D-printing technology.

Chapter 10: Nanotechnology in Africa: This chapter discusses the use of nanotechnology in Africa. Nanotechnology is the science and technology of manipulating a matter at the nanoscale. It is a disruptive emerging technology dedicated to the study and manipulation of characteristics of matter at the atomic level. Africa is lagging behind other continents in terms of nanotechnology research, inventions, standards, and the number of companies operating in that area. Ultimately, nanotechnology will profoundly affect Africa's economy, regardless of its level of direct participation.

This book is a comprehensive text on the use of emerging technologies in various nations of Africa. It provides an overview of each technology, its applications, and adoption in Africa so that beginners can understand it, its increasing importance, and its relevance. It is a must read for those interested in the socio-economic development of Africa.

I am grateful for the support of Dr. Annamalia Annamalai, Head of the Department of Electrical and Computer Engineering, and Dr. Pamela Obiomon, Dean of the College of Engineering at Prairie View A&M University, Prairie View, Texas. Special thanks are due to my wife Dr. Janet Sadiku for helping in various ways.

– M. N. O. Sadiku

Contents

About the Author

Matthew N. O. Sadiku received his B.Sc. degree in 1978 from Ahmadu Bello University, Zaria, Nigeria, and M.Sc. and Ph.D. degrees from Tennessee Technological University, Cookeville, TN, in 1982 and 1984, respectively. From 1984 to 1988, he was an Assistant Professor at Florida Atlantic University, Boca Raton, FL, where he did graduate work in computer science. In total, he received seven college degrees. From 1988 to 2000, he was at Temple University, Philadelphia, PA, where he became a Full Professor. From 2000 to 2002, he was with Lucent/Avaya, Holmdel, NJ, as a system engineer and with Boeing Satellite Systems, Los Angeles, CA, as a senior scientist. He is presently a Regents Professor Emeritus of electrical and computer engineering at Prairie View A&M University, Prairie View, TX.

He is the author of over 1,430 professional papers and over 150 books, including *Elements of Electromagnetics* (Oxford University Press, 7th ed., 2018), *Fundamentals of Electric Circuits* (McGraw-Hill, 7th ed., 2020, with C. Alexander), *Computational Electromagnetics with MATLAB* (CRC Press, 4th ed., 2019), *Principles of Modern Communication Systems* (Cambridge University Press, 2017, with S. O. Agbo), and *Emerging Internet-Based Technologies* (CRC Press, 2019). In addition to the engineering books, he has written Christian books including *Secrets of Successful Marriages*, *How to Discover God's Will for Your Life*, and commentaries on all the books of the New Testament Bible. Some of his books have been translated into French, Korean, Chinese (and Chinese Long Form in Taiwan), Italian, Portuguese, Spanish, German, Dutch, Polish, and Russian.

He was the recipient of the 2000 McGraw-Hill/Jacob Millman Award for outstanding contributions in the field of electrical engineering. He was also the recipient of Regents Professor Award for 2012–2013 from the Texas A&M University System. He is a registered professional engineer and a life fellow of the Institute of Electrical and Electronics Engineers (IEEE) "for contributions to computational electromagnetics and engineering education." He was the IEEE Region 2 Student Activities Committee Chairman. He was an associate editor for *IEEE Transactions on Education*. He is also a member of Association for Computing Machinery (ACM). His current research interests are in the areas of computational electromagnetic, computer science/networks, engineering education, and marriage counseling. His works can be found in his autobiography, *My Life and Work* (Trafford Publishing, 2017) or his website: www.matthew-sadiku.com. He can be reached via email at sadiku@ieee.org

https://doi.org/10.1515/9783112211984-204

Chapter 1
Introduction

The only thing predictable about Africa is its unpredictability. – Brian Jackman

1.1 Introduction

Africa is the second largest continent in the world after Asia and the second largest continent in population. More than 1.3 billion people live on the African continent, which implies that about 15% of the world's total population live in Africa. Many scientists consider Africa to be the origin of mankind with cultural diversity. It is home to incredible wildlife and all kinds of unique, healthy, and delicious foods. It is a continent rich in resources, especially food-related products. As multiple nations and cultures make up Africa, the cuisine is each nation changes considerably. There are 54 countries in Africa: Algeria, Angola, Benin, Botswana, Burkina Faso, Burundi, Cameroon, Cape Verde, Central African Republic, Chad, Democratic Republic of the Congo, Djibouti, Egypt, Equatorial Guinea, Eritrea, Ethiopia, Gabon, Gambia, Ghana, Guinea Bissau, Guinea, Ivory Coast, Kenya, Lesotho, Liberia, Libya, Madagascar, Malawi, Mali, Mauritania, Mauritius, Morocco, Mozambique, Namibia, Niger, Nigeria, Republic of the Congo, Reunion, Rwanda, Senegal, Seychelles, Sierra Leone, Sao Tome and Principe, Somalia, South Africa, Sudan, Swaziland, Tanzania, Togo, Tunisia, Uganda, Western Sahara, Zambia, and Zimbabwe [1, 2]. As shown in Figure 1.1, the huge continent is divided into five regions: Northern Africa, Eastern Africa, Central Africa, Southern Africa, and Western Africa [3].

Africa is a continent that is rapidly embracing cutting-edge technology to leapfrog into the future. The continent is not just catching up with the world; it is propelling itself to the forefront of innovation. Africa is rising, and its tech scene is leading the way [4]. Africa is closely watched as the next big growth market. Africa is home to a burgeoning pool of skilled professionals in the fields of technology, engineering, medicine, and data science.

In spite of these reasons for optimism, the promise remains unfulfilled. Africa's diverse populations are still often misrepresented and misunderstood on the global stage, an insidious legacy of the colonial gaze. Scholars often regard Africa as the less developed economies characterized by low rates of technological adoption, a huge gap in the digital divide, low levels of industrialization, and political and cultural marginalization. Although these difficulties weaken the capacity of the African nations to attain the Sustainable Development Goals (SDGs), they can be addressed with the use of emerging technologies. Emerging technologies can help Africa improve industrialization, thereby increasing the economic growth and development [5].

https://doi.org/10.1515/9783112211984-001

Figure 1.1: The map of Africa showing five subregions [3].

Technologies are unlocking new pathways for rapid economic growth, innovation, job creation, and access to services in Africa. Emerging technologies promise to reduce the costs of undertaking research, offer endless opportunities for innovations and present opportunities to catch up with leading nations. These technologies can enable economic transformation in Africa and help create more jobs for its people. They are transforming all elements of the continent's economy, from education to healthcare, agriculture to telecommunications [6].

This chapter highlights the opportunities and challenges of implementing emerging technologies in Africa. It unpacks what some of the economies in Africa have achieved in terms of emerging technologies. It begins with describing what emerging technologies are and explains some of them. It covers how African nations are adapting emerging technologies. It highlights the benefits and challenges of emerging technologies in Africa. The last section concludes with comments.

1.2 Emerging Technologies

Our society relies heavily on technology. Technology is everywhere. It surrounds every aspect of twenty-first century life. It is in the cell phones we use, the cars we drive, and even the food we eat. Technology connects and brings us together to share ideas across borders and cultures. As digital technology has become ubiquitous, affordable, and portable, an increasing number of people worldwide are increasing their online participation. People use different technologies and tools for different reasons, such as exchanging messages, meeting, entertainment, shopping, sharing photos, and communicating with people. The use of technology has transformed every discipline and career, from engineering to medicine to politics. Technology is a star attraction for the whole world in the new age.

Technology is essentially the organized application of knowledge to solve practical problems. It refers to the application of knowledge for practical purposes and for human benefits. It is also a collection of systems designed to perform some function. Emerging technology is a general term used to describe new technology. These are those technologies that are at the beginning stages of development, where they are beginning to emerge into popularity. They refer to technologies that are still being developed and play a key role in industrial modernization. They play a key role in industrial modernization and are used by both the private and public sectors. Emerging technologies have revolutionized education, manufacturing, healthcare, government, business, the military, etc. [7]. They have proven useful and fruitful in developed economies, especially in improving efficiency in the industrial and service sectors, as well as improving economic growth.

To be defined as emerging, technology must have quick growth, influence, revolutionary originality, and consistency. Some of the characteristics of emerging technologies are [8, 9]:
1. they may or may not be new technologies;
2. they change rapidly, so they are always in a state of coming into being;
3. they go through cycles of hyped expectations;
4. they are in a continuous state of being understood and researched; and
5. they have the potential to transform social practices.

Here are the top emerging technologies that are currently transforming the economy globally [10]:
1. Artificial intelligence (AI)
2. Machine learning
3. Robotic process automation
4. Three-dimensional (3D) printing (3DP)
5. Cloud computing
6. Augmented/virtual reality
7. Nanotechnology

8. Internet of things (IoT)
9. Blockchain
10. Drones
11. Big data
12. 5G network

These emerging technologies have literally transformed modern agriculture, education, healthcare, manufacturing, and defense. They play an important role in improving business efficiency, industry operations, and trade. Government, universities, public and private sectors including large enterprises, small and mid-size enterprises (SMEs) and start-ups play a dominant role in the innovation, development and application of emerging technologies. There has been discussion on "Talking Up Africa" episodes around the transformative power of emerging technologies, particularly AI. The African Union has prioritized various emerging technologies with potential to improve health on the continent.

1.3 Use of Emerging Technologies

Emerging technologies such as CCTV cameras with facial recognition systems, drones, AI, robots, blockchain, and "smart cities" are proliferating in Africa. They are having a powerful impact on the security and stability of African states. Digitization is improving the government revenue collection and curbing corruption. Drones are delivering life-saving medical supplies. Today, Africa's mobile service subscription figures are skyrocketing. Yet with each advance there is a cost [11]. Access to technology must be coupled to socioeconomic welfare. We consider the impact of the following emerging technologies are making on Africa:

– *AI:* The umbrella term of "artificial intelligence" refers to a set of tools and capabilities like machine learning, expert systems, robotics, and natural language processing. AI is a branch of computer science that studies and develops intelligent machines. It is becoming more increasingly popular and is covering all facets of human activity. AI harnesses big data, using machine learning to make predictions and decisions. AI can be used in the production of pharmaceutical products, especially locally in African economies, to reduce the volume imported from other economies. AI creation and implementation are transforming lives and cultures in a variety of ways including economically, socially, and politically. The application of AI in education, healthcare, agriculture, commerce, and governance is showing a significant impact on the various sectors. For example, Chatbots in Kenya now provide healthcare services to people without visiting doctors. Mama Money and Mukuru enable easy and quick transfer of money across different countries in Africa. AI is a game-changing innovation with the potential to improve all sectors of the African social system. However, the adoption and use of AI applications in African community raise some issues including skills acquisition, ethics, programming, data integration, user attitude, government policy, and insufficient infrastructure and network connectivity. The knowledge of AI is still at an early stage and the population is still not certain of the advantages in developing African continent. There is a need for a policy on AI implementation strategies in African countries [12]. US technology multinationals are investing in AI and other emerging technologies because they recognize the potential impact these technologies can have on global health. AI has the potential to reshape economies and societies, and it is already making waves in Africa.

– *3DP:* 3DP (also known as additive manufacturing or rapid prototyping) was invented in the early 1980s by Charles Hull, who is regarded as the father of 3DP. Since then, it has been used in manufacturing, automotive, electronics, aviation, aerospace, consumer products, education, entertainment, medicine, space missions, the military, and chemical and jewelry industries. A 3D printer works by "printing" objects. The manufacturing process creates parts layer by layer until the entire 3D part is complete. The technology has been transforming all industries such as manufacturing, construction, fashion, jewelry, sculpture, architecture, and aerospace and by producing 3D objects based on the commands given by software programs. Instead of using ink, it uses more substantive materials – plastics, metal, rubber, and the like. It scans an object – or takes an existing scan of an object – and slices it into layers, which can then convert into a physical object. Layer by layer, the 3D printer can replicate images created in computer-aided design (CAD) programs. In other words, 3DP instructs a computer to apply layer upon layer of a specific material (such as plastic or metal) until the final product is built. This is distinct from conventional manufacturing methods, which often rely on removal (by cutting, drilling, chopping, grinding, forging, etc.) instead of addition. Models can be multi-colored to highlight important features, such as tumors, cavities, and vascular tracks. 3DP technology can build a 3D object in almost any shape imaginable as defined in a CAD file. This emerging technology can be used in the industrial sector for manufacturing purposes in Africa.

– *Blockchain:* Blockchain (also known as "distributed ledger technology") is a peer-to-peer network that sits on top of the Internet. Blockchain technology is an innovation which is regarded as the center of Industry 4.0 revolution and it has become part of our lives. It is a system that stores data in a special way. Blockchain technology has some interesting properties, such as its decentralized nature, immutability, decentralization, transparency, and permissionless, that may be used to address pressing issues in many sectors. Although this technology finds its first application in the financial sector, it has become possible to use it in all sectors which can be integrated with technology today. Before blockchain technology, people turned to gold or real estate when inflation hit its peak. Today, governments all over the globe have started opening up to blockchain and crypto. By using blockchain, governments can reduce administrative costs, increase transparency, and improve service delivery. Blockchain is revolutionizing the digital world by bringing a new perspective to security, efficiency, and stability of systems and data. It is network of computers that is decentralized. Blockchain keeps a track of distributed data and provides encrypted transaction tracking. It has attracted attention with its unique characteristics, such as irrevocability and security. It will be a part of our everyday life. Blockchain can be described as a shared digital record of transactions which can be verified by a network of participants. It can accommodate self-executing contracts commonly referred to as smart contracts. It has the potential to revolutionize the way multiple industries operate in a country. The decentralized nature of blockchain allows for the integration and distribution of transactional data from multiple sources using cloud services. It also ensures data integrity by providing a single source, eliminating duplication and strengthening security. This technology is of great value and is highly sought after and many industries such as manufacturing are ready to implement it, resulting in widespread benefits for the population. Blockchain could be a bedrock of economic innovation and growth in Nigeria.

– *Cloud Computing:* Cloud technology is one of the globally recognized emerging technologies in the new millennium that are most likely to change people's lives. Organizations with not enough resources to build their own infrastructure can now take advantage of the cloud services to suit their specific needs. Rather than building, owning, and maintaining their own IT infrastructure, businesses can use cloud to access technology resources such as computing capacity, storage, and databases on a pay-as-you-go basis. An industry needs the cloud for the following reasons: (1) Mobile workforce: empowering employees to sift real-time data and make decisions on the fly; (2) minimize disruptions: with the right sort of cloud setup problems can be anticipated and solved quickly; (3) collaboration: with the right technology, collaboration – as well as transparency and accountability – are easily managed; (4) innovation: product innovation and process innovation are powerful weapons to survive or thrive in such an environment; and (5) lower cost: no hardware procurement, maintenance, or staff is needed to operate the systems. Cloud computing is essentially accessing computing services through the Internet. A cloud platform makes it easier for students and teachers to access educational resources. Amazon Web Services plans to set up its cloud infrastructure in Kenya, bringing cloud computing, storage, database, and other services closer to end customers and on-premises data centers.

– *Robotics:* Robotics constitutes one of the most exciting fields of technology and rapidly growing field today. It is the discipline of designing and constructing intelligent machines, called robots. A robot is an autonomous mechanical device that is designed to sense its environment, carry out computations to make decisions, and perform actions like humans in the real world. Popular interest in robotics has increased in recent years. Robots are becoming more and more common in our society and more integrated into our lives. This is due to the fact that they are becoming smarter, smaller, cheaper, faster, more flexible, and more autonomous than ever before. Robotics has advanced and taken many forms including fixed robots, collaborative robots, mobile robots, industrial robots, medical robots, police robots, military robots, officer robots, service robots, space robots, social robots, personal robots, and rehabilitation robots [13, 14]. Robotics technology has been implemented in a variety of fields including manufacturing, medicine, elderly care, rehabilitation, education, agriculture, home appliances, search and rescue, car industry, defense, and more. Robotics is an interdisciplinary discipline embracing mechanical engineering, electrical engineering, computer science, and others. The goal of robotics is to create intelligent machines (called robots) that behave and think like humans. Robots were originally intended for use in industrial environments to replace humans in tedious and repetitive tasks. Today, robots help human beings in everyday life. They are regarded as intelligent agents that can perform actions similar to what humans can do. This is one of the most exciting Educational robotics helps to enhance strategic problem-solving, higher-order thinking, logical and analytical reasoning, computational thinking, teamwork, collaborative skills, and more. For example, educational service robots have appeared in the United States, Canada, Japan, South Korea, and Taiwan. Robotic process automation (RPA) is a type of software designed to automate different business tasks such as transaction processing, customer service calls, and data management. The goal of RPA is to make work easier by delivering the desired outcomes.

– *5G Network:* This represents more than just another step in the evolution of wireless technologies. It is the convergence of wireless with computing and the cloud. 5G will enable applications that are especially relevant for African countries such as communications, agriculture, healthcare, education, mining, manufacturing, public safety, and disaster response. It is widely understood that connectivity is a key factor for success in the increasingly digital economy, and Africa has a great chance to leverage the potential of this emerging technology. African Administrations should define and agree on a 5G plan and implementation timeline aimed at achieving coordinated and harmonized regional deployment. It was also recommended that African Administrations should assign spectrum for 5G in low, mid, and high bands in sufficient quantities to support 5G rollout [15].

– *The IoT:* The IoT refers to the billions of physical devices connected to the wireless Internet that allows exchanging data around the world. It is a global infrastructure for the information society, connecting devices/things to the Internet and to each other using wired or wireless technology. The devices include smart phones, tablets, desktop computers, autonomous vehicles, refrigerators, toasters, thermostats, cameras, pet monitors, alarm systems, insulin pumps, industrial machines, intelligent wheelchairs, wireless sensors, mobile robots, etc. The IoT connects people, places, and products, offering new opportunities to generate value in products and business processes. With a growing internet penetration of about 60%, more people in Africa are connected than ever before. Additionally, sensor, drone, and satellite technologies can collect demographic, climate, agricultural, and transportation data to support green growth financing strategies. Data hubs leveraging IoT can help break data silos, with data collected across agencies, the public and private sectors, and countries and regions. Of course, African governments must have strong data protection and privacy policies to ensure that such data is only used ethically [16].

1.4 African Nations Adapting Emerging Technologies

Government, universities, public, and private sectors have an important role to play with diffusion of emerging technology on the supply side, and adoption and usage of the technology on the demand side. The adoption and use of the emerging technologies in the African context are currently low due to some emerging challenges. African nations are well-positioned to take advantage of lessons learned from other nations and can leverage their strengths. Although mobile internet availability has increased, Africa's internet penetration still lags behind other continents – with digital divides still an issue in remote areas in all African nations. The African continent has become a dumping ground for new technologies. Sectors such as agriculture, health, finance, and energy are being upturned by a wave of automation, AI, biotechnology, big data, synthetic biology, and geoengineering. Most African national governments are already making efforts, through a number of initiatives, to harmonizing emerging technologies with the contemporary public service sector in a way that strengthens the efficiency of the sector. Here we examine the following selected African nations to see how they are adopting emerging technologies [5, 17]:

– *South Africa:* Africa's computing history dates back to 1921 when South Africa took delivery of its first tabulating equipment from the then Computing-Tabulating-Recording Company which later became IBM. South Africa exhibits technological leadership in the areas of mobile software, security software, and online banking. Furthermore, the South African government targets a massive skills development initiative through various programs and agencies with the goal to educate a million young people in robotics, AI, coding, cloud computing, and networking by 2030. This target may not be achieved due to financial constraints. The nation should tap into emerging technologies to reduce cybercrime.

– *Zimbabwe:* In 2020, only 27% of the total population in Zimbabwe had access to the Internet and can participate in e-commerce activities. Poor information and communication technology (ICT) infrastructure, poor ICT policies, technological affordability, and technological competence limit Zimbabweans' ability to enjoy ICT benefits. However, Zimbabwe has adopted AI technology in the banking sector. At the grassroots level, Zimbabwe introduced computer programs at primary schools in different provinces. Some higher learning institutions have developed e-learning platforms, while others are still lagging.

– *Kenya:* This is home to what is known as a "Silicon Savanah." Kenya has a growing tech-savvy ecosystem. Over 70% of Kenyans have a mobile money account in M-Pesa, and over 75% of Kenyans aged 15 or older made a mobile payment in the last year. During the past decade, Kenya has advanced quickly as a hotspot for some of the continent's most innovative digital enterprises. Kenya's policymakers have enabled a favorable regulatory environment, and have promoted a high use of digital payments. In Kenya, 4G networks are well-established and 5G is being tested for its capabilities in different ways. Safaricom, Kenya's leading telecommunications company has announced that 5G will be fully operational in the country soon. Kenya's currency is shown in Figure 1.2 [18].

Figure 1.2: Kenya's currency [18].

– *Rwanda:* Rwanda has been moving to transform itself into a digital hub, with several notable initiatives. Rwanda's Mara Group became the first manufacturers of a smartphone made entirely in Africa. Also, Rwanda has played a pioneering role in the region in exploring several key emerging technologies, such as drones used to deliver critical supplies to inaccessible areas or considering a central bank issued digital currency. Figure 1.3 shows Airtel Africa's Group CEO sealing deal with Rwanda [19].

– *Egypt:* The digital technology sector is Egypt's second-fastest growing sector. Egypt is a regional leader in skilled digital jobs creation with online freelancer pools in creative and multimedia, software development and technology, and in writing and translation. Egypt is also developing one of the region's fastest growing entrepreneurial hubs.

– *Nigeria:* Nigeria has a powerful entrepreneurial climate, with innovative ventures. These ventures cut across the education, fintech, agriculture, healthcare, logistics, and

Figure 1.3: Airtel Africa's Group CEO sealing deal with Rwanda [19].

travel. Lagos' Yaba neighborhood has even earned the nickname "Yabacon Valley." A unique identity system is essential in developing countries, where the vast majority have few other ways to prove who they are and thereby get access to public services or the financial system, usually through a mobile phone.

– *Ethiopia:* Ethiopia is experiencing positive developments in several areas that can facilitate digitally enabled growth. Ethiopia has also been upgrading its infrastructure, with a $20 billion investment in the power sector. Overall enrollment in higher education facilities in the country has grown fivefold since 2005, and the government has a policy of training 70% of students in STEM. With a fast-emerging tech hub, also known as "Sheba Valley," the country has had several homegrown ride-hail ventures. There is a growing manufacturing industry and use of advanced technologies, such as blockchain use in tracking the supply chain.

The digital advantages and gaps of different countries vary widely. Figure 1.4 shows how six African countries compare on nine metrics needed for digital growth [17].

1.5 Benefits

Emerging technologies such as AI, blockchain, and the IoT have the capability to provide firms with major technological advantages in the coming years. They have the potential to accelerate the economic transformation of developing countries. They have also helped many African countries overcome some pertinent issues, such as corruption and efficient agriculture. Other benefits of emerging technologies in Africa include the following:

Egypt   Ethiopia   Kenya   Rwanda   Nigeria   South Africa

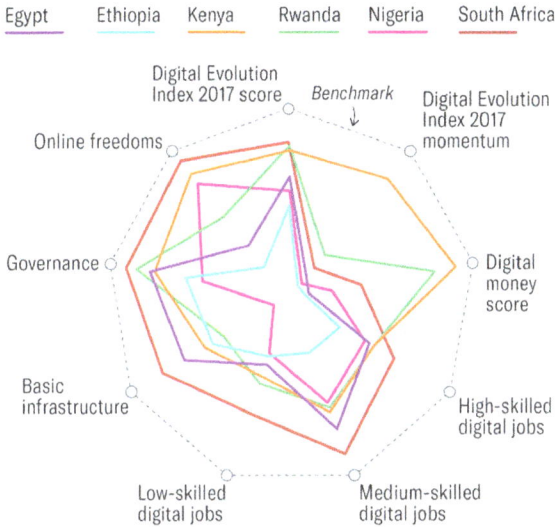

Figure 1.4: How six African countries compare on nine metrics needed for digital growth [17].

1. *Interest in Technology:* Africans are not anti-technologies. Rather they are for technologies that do not jeopardize their interests as individuals or continent. Africa has not disappointed in delivering innovative emerging technologies through these start-ups, with successful ventures coming from Nigeria, Kenya, and South Africa in sectors varying from fintech, to agritech, to healthcare. Africans are harnessing the opportunities available to build solutions with emerging technologies to implement into several sectors. GMOs should not be part of our food system because it would negatively affect our health and biodiversity.

2. *SDGs:* Various sectors such as business, manufacturing, and energy can all benefit from the use of emerging technologies. The long-term benefits of harnessing emerging technologies in Africa will be the attainment of some of the SDGs, such as poverty reduction, increased economic growth, improved industrialization levels, and clean energy.

3. *Prospects:* Emerging technologies present prospects for the development of new markets, the provision of new services, the expansion of current manufacturing processes, job creation, the generation of additional income, and the mitigation of environmental damage. They transform social practices, improve economic growth, and create jobs. Creating meaningful employment opportunities for Africa's youth is already a major development policy issue.

4. *Data Democratization:* Cypher Crescent in Nigeria highlighted the importance of data democratization and digitalization. Their technology helps companies integrate data to create a single source of truth for informed decision-making. The impact of technology extends across various sectors, including healthcare and finance.

5. *Financial Inclusion:* In a review of the impact of financial inclusion on economic growth, the World Bank argues that "such services must be provided responsibly and safely to the consumer, and sustainably to the provider." Financial inclusion has the potential to reduce poverty and inequality by helping disadvantaged groups to benefit from opportunities that otherwise would not have been available. Innovation in financial services through time has expanded access to and improved financial inclusion globally. There is a need to provide improved systems for poverty reduction, if not alleviation.

1.6 Challenges

There is serious experimentation around emerging technologies happening in the background in Africa. Some economies in the Africa face problems, such as digital illiteracy, the energy crisis, high poverty rates, unemployment, adverse climate change, poor education, poor health facilities, insecurity, corruption, bad leadership, lack of political will, lack of safe water and sanitation, inefficiency in the industrial sector, user attitude, and trade deficits. Also, associated challenges of emerging technologies often lead to abandoned solutions in the early stages. Unfortunately, it is still difficult for some political and business leaders to quantify the benefits associated with the emerging technologies. Most of the African population take a "wait-and-see" approach to technology adoption. Other challenges of emerging technologies in Africa include the following [12, 20, 21]:

1. *Education Gap:* One of the most powerful instruments that can be used for reducing any sort of inequality, reducing poverty, and laying a foundation for a sustained economy is education. Lack of education is one of the leading reasons Africans are lagging behind the rest of the world. Inequality in Africa has always been there and currently, it is on the rise, especially in urban areas. The education levels have been further hampered by the prevailing unhealthy macroeconomic environment created in the last two decades or so.

2. *Poor Infrastructure:* One of the primary challenges is the lack of infrastructure. Many African countries, including Zambia, have limited access to electricity and Internet connectivity. The poor infrastructure in Africa is also a contributing factor to why African nations are lagging behind. In Africa, the connectivity is very less as compared to the rest of the world. Another dividing factor in African countries is accessibility to electricity. In most of the countries of Africa, only a slight percentage more than half of the households have access to electricity and in some countries such as Tanzania, Uganda, Rwanda, and Ethiopia have electricity available to less than 20% of the households. Due to the unavailability of electricity, the African people are not able to enjoy the latest technology as the rest of the world.

3. *Digital Illiteracy:* This refers to the inability to use effectively digital media such as a computer, software, tablet, smartphone, or the Internet. It is the absence of information, skills, or comprehension regarding the utilization of digital tools. This lack of literacy can prohibit persons from using a computer, smartphone, or other electronic devices. It means not having the skills to explore, create, and manage digital content. Digital illiteracy is a major stumbling block that affects economies in Africa's ability to exploit emerging technology opportunities. Digital illiteracy and a poor ICT infrastructure affect many African economies, such as South Africa, Zimbabwe, Burundi, Somalia, Malawi, Angola, Liberia, Cameroon, Chad, and Comoros.

4. *Corruption:* Another leading factor to why Africa is lagging behind the rest of the world in innovating technology is corruption. As Dennis Prager rightly said, "Corruption is Africa's greatest problem. Not poverty. Not lack of riches, No racism." A corrupt government leads to poor education, a poor healthcare system, poverty, the economic gap between classes of people, economic instability, malnutrition, non-existent security, etc.

5. *Cost:* With advance, there is a cost. We must bear in mind that these emerging technologies will come at a cost to the community at large. The cost will include loss of jobs due to automation, amplifying bias, increasing social surveillance, and the high environmental costs of computing power. Financial resources play a key role in the acquisition of emerging technologies.

6. *Lack of Internet Access:* Digital access is still highly unequal among African nations. Higher prices of Internet as well as illicit financial flows pose a serious difficulty for African economies to harness emerging technologies. Most Africans have no Internet access because they cannot afford it. Large-scale investment in infrastructure is crucial for building an inclusive digital future.

7. *Digital Divide:* There is a digital divide between the poor and the rich. Without massive investment in infrastructure, this digital divide will continue to widen. Less developed regions risk being left behind. Nigerian start-up uLesson is helping bridge network connectivity gaps and addressing high data costs by offering a non-streaming option for their pre-recorded secondary school education content

8. *Bridging the Gap:* One may concern raised during the "Talking Up Africa" episode is the disparity in AI development across Africa. While countries like South Africa and Kenya are making strides, Francophone Africa is lagging behind. Bridging this gap and ensuring equitable AI development is a priority for African policymakers.

9. *Skills Shortage:* There is a global skill shortage about AI. The majority of AI expertise resides in the United States, Europe, and China due to significant investments in these regions. This shortage calls for strategic thinking from policymakers to cultivate AI talent and strategies for its widespread adoption. AI should be taught in African universities and colleges.

10. *Collaboration:* It is so important to cooperate on a global scale. All countries have something to offer, and all deserve a seat at the table. Technological progress is often presented as a race or competition, where the most powerful countries and companies vye for total domination. But to truly harness the power of the Fourth Industrial Revolution, we need all countries to collaborate. If we succeed in pooling our collective ingenuity, creativity, resources, and ambitions, we can build a world that is truly prosperous for all.

11. *No Regulation:* Due to the weak or no regulatory systems in African countries, corporations use Africa as a petri dish for technological adventurism while making her the dumping ground for unwanted technological mistakes.

These challenges may have a direct influence on African economic development. Without addressing the challenges, Africa will continue to lag behind other continents in the global world in using emerging technologies.

1.7 Conclusion

Emerging technologies have been described as radically novel and relatively fast-growing technologies whose deployment and utility may incur disruptive effects in all sectors of the economy or societies. Emerging technologies play an important role in improving business efficiency, industry operations, and trade. The crucial factors needed for technology adoption are sadly lacking across most of African nations. African economies need develop sound and robust ICT policies that match the digital needs of their communities. African governments should prioritize Afro-centric R&D investments, focusing on producing and commercializing scientific knowledge.

African countries that take advantage of the emerging technologies and limit the risks inherent in them may achieve greater peace and prosperity. African countries cannot afford to be left behind. Addressing critical societal issues will remain at the forefront of Africa's tech agenda. Africa is becoming a hub for the development and application of AI and other emerging technologies. More information about the use of emerging technologies can be found in books [22–24].

References

[1]   "27 surprising facts about Africa," https://www.signatureafricansafaris.com/africa-facts/
[2]   M. N. O. Sadiku and J. O. Sadiku, *Traditional Foods Around the World*. Moldova, Europe: Lambert Academic Publishing, 2023, p. 162.
[3]   "Regions of Africa," https://www.worldatlas.com/geography/regions-of-africa.html
[4]   "The top 5 emerging tech trends in Africa," https://www.linkedin.com/pulse/top-5-emerging-tech-trends-africa-nerdzfactoryorg

[5] "Opportunities in emerging technologies for Southern Africa: How the Global South should adopt to take advantage?" https://onlinelibrary.wiley.com/doi/full/10.1002/isd2.12321#:~:text=In%20terms%20of%20policy%20recommendations,invest%20in%20research%20and%20development.

[6] M. N. O. Sadiku, J. O. Sadiku, and U. C. Chukwu, "Use of emerging technologies in Africa," *Proceedings of the 2024 International Conference on Scientific Computing*, Las Vegas, 2024.

[7] M. N. O. Sadiku and O. D. Olaleye, *Emerging Technologies in Education*. Metairie, LA: The Ewings Publishing, 2024, pp. 1–25.

[8] V. Bozalek, D. Ng'ambi, and D. Gachago, "Transforming teaching with emerging technologies: Implications for higher education institutions," https://core.ac.uk/download/pdf/83123732.pdf

[9] G. Veletsianos, "A definition of emerging technologies for education," in G. Veletsianos (ed.), *Emerging Technologies in Distance Education*. Athabasca University Press, Chapter 1, ·2010, pp. 3–22.

[10] "Understanding the value and impact of emerging technologies," https://techtrends.africa/understanding-the-value-and-impact-of-emerging-technologies/

[11] N. Allen, "The promises and perils of Africa's digital revolution," March 2021, https://www.brookings.edu/articles/the-promises-and-perils-of-africas-digital-revolution/

[12] A. Ade-Ibijola and C. Okonkwo, "Artificial intelligence in Africa: Emerging challenges," in D. O. Eke, K. Wakunuma, and S. Akintoye (eds.) *Responsible AI in Africa: Social and Cultural Studies of Robots and AI*. Palgrave Macmillan, 2023, pp. 101–117.

[13] R. D. Davenport, "Robotics," in W. C. Mann (ed.), *Smart Technology for Aging, Disability, and Independence*. John Wiley & Sons, 2005, Chapter 3, pp. 67–109.

[14] M. N. O. Sadiku, S. Alam, and S. M. Musa, "Intelligent robotics and applications," *International Journal of Trends in Research and Development*, vol. 5, no. 1, January-February 2018, pp. 101–103.

[15] "Enabling of 5G and other emerging ICT technologies in Africa; discussions and recommendations," https://atuuat.africa/enabling-of-5g-and-other-emerging-ict-technologies-in-africa-discussions-and-recommendations/

[16] B. Bayou and R. Floyd, "How emerging technology can boost Africa's green industrial future," https://acetforafrica.org/research-and-analysis/insights-ideas/commentary/how-emerging-technology-can-boost-africas-green-industrial-future/#:~:text=Africa's%20climate%2Dpositive%20industrial%20transformation,information%20into%20forecasts%20and%20simulations.

[17] B. Chakravorti and R. S. Chaturvedi, "Research: How technology could promote growth in 6 African countries," https://hbr.org/2019/12/research-how-technology-could-promote-growth-in-6-african-countries

[18] "Browsing: Emerging technologies in Africa," Unknown Source

[19] A. Turner, "Emerging technologies can drive Africa's digital and financial inclusion," https://www.mobileeurope.co.uk/emerging-technologies-can-drive-africas-digital-and-financial-inclusion/

[20] "Exploring the impact of emerging technologies in Africa: A recap of 'Talking Up Africa' episode 4," https://www.linkedin.com/pulse/exploring-impact-emerging-technologies-africa-recap#:~:text=episode%204%20underscored%20the%20vast,sectors%2C%20including%20healthcare%20and%20finance.

[21] "4 Reasons why Africa is lagging behind the world in technology," https://allafrica.com/stories/202109140845.html.

[22] National Research Council, et al. *Emerging Technologies to Benefit Farmers in sub-Saharan Africa and South Asia*. National Academies Press, 2009.

[23] M. Masinde and A. Bagula (eds.), *Emerging Technologies for Developing Countries: 5th EAI International Conference, AFRICATEK 2022, Bloemfontein, South Africa, December 5–7, 2022, . . . and Telecommunications Engineering)*. Springer, 2023.

[24] B. Kastel, *Techies, Trust, and Trillionaires: How Blockchain and Emerging Technologies Create Money for Africa*. Competitive Edge International, 2018.

Chapter 2
Articifial Intelligence in Africa

Artificial intelligence will digitally disrupt all industries. Don't be left behind. – Dave Waters

2.1 Introduction

Africa is the second largest continent in the world after Asia and the second largest continent in population. More than 1.3 billion people live on the African continent, which implies that about 15% of the world's total population lives in Africa. Africa is a continent that is rapidly embracing cutting-edge technology to leapfrog into the future. Africa is underrepresented in artificial intelligence (AI). The continent is not just catching up with the world; it is propelling itself to the forefront of innovation. Africa is rising, and its tech scene is leading the way [1]. Africa is closely watched as the next big growth market. It is the home to some of the youngest populations in the world. The African Union (AU, made up of 55 member nations) is preparing an ambitious AI policy that envisions an Africa-centric path for the development and regulation of this emerging technology. The AU is a regional body designed to guide the prosperity at a continental level and as such is expected to play a significant role in the adoption of AI. The heads of African governments are expected to eventually endorse the continental AI strategy. Time is not with Africa and urgent interventions are required. The power of AI starts with people and intelligent technologies working to create better outcomes for customers and society [2].

Although AI is a branch of computer science, there is hardly any field which is unaffected by this technology. Common areas of applications include agriculture, business, law enforcement, oil and gas, banking and finance, education, transportation, healthcare, engineering, automobiles, entertainment, manufacturing, speech and text recognition, facial analysis, telecommunications, and the military. AI has endless potential to handle tasks commonly done by humans, including natural language processing (NLP), image recognition and data analytics, visual perception, decision-making, speech recognition, business process management, and even the diagnosis of disease, all of which normally require human intelligence [3]. Today, AI is integrated into our daily lives in several forms, such as personal assistants, automated mass transportation, aviation, computer gaming, facial recognition at passport control, voice recognition on virtual assistants, driverless cars, companion robots, etc. AI will push us to rethink the social contract at the heart of our democracies, our education models, labor markets, and the way we conduct warfare. The inherent nature of AI is without doubt a threat to the rule of law. But the blame for the erosion of the rule of law cannot be put squarely at the foot of technology [4].

https://doi.org/10.1515/9783112211984-002

AI, a fast-evolving technology that taps the intelligence of machines, is transforming all social spheres globally. AI for Africa presents opportunities to put the continent at the forefront of the Fourth Industrial Revolution. Adopting and implementing AI in Africa could be significant if the focus becomes Africa-centered, nurturing local solutions to local challenges. It is essential that AI is adapted to the continent's interests, values, and cultures. Africa needs supportive policies and robust infrastructure to tap the limitless opportunities of AI. Africa can jump on the bandwagon of AI if they further unleash the potential of its young population in terms of innovation, creativity, and discovery [5].

This chapter examines the adoption of AI in African nations. It begins with explaining what AI is all about and covers its major components. It discusses the use and adoption of AI in Africa. It covers African nations adapting AI. It highlights the benefits and challenges of using AI in Africa. The last section concludes with comments.

2.2 What Is Artificial Intelligence?

The term "artificial intelligence" (AI) was first used at a Dartmouth College conference in 1952. AI refers to the ability of a computer system to perform human tasks (such as thinking and learning) that usually can only be accomplished using human intelligence [6].

AI, sometimes called machine intelligence, refers to intelligence demonstrated by machines, while the natural intelligence is the intelligence displayed by humans and animals. AI is an umbrella term John McCarthy, a computer scientist, coined in 1955 and defined as "the science and engineering of intelligent machines." AI is now the latest big game changer. Typically, AI systems demonstrate at least some of the following human behaviors: planning, learning, reasoning, problem solving, knowledge representation, perception, speech recognition, decision-making, language translation, motion, manipulation, intelligence, and creativity. AI is an interdisciplinary and comprehensive field covering numerous areas such as computer science, psychology, linguistics, philosophy, neurosciences, cognitive science, thinking science, information science, system science, and biological science. Today, AI is integrated into our daily lives in several forms, such as personal assistants, automated mass transportation, aviation, computer gaming, facial recognition at passport control, voice recognition on virtual assistants, driverless cars, and companion robots.

Although AI is a branch of computer science, there is hardly any field which is unaffected by this technology. Common areas of applications include agriculture, business, law enforcement, oil and gas, banking and finance, education, transportation, healthcare, engineering, automobiles, entertainment, manufacturing, speech and text recognition, facial analysis, telecommunications, and military. AI has endless potential to handle tasks commonly done by humans, including NLP, image recognition and data analytics, visual perception, decision-making, speech recognition, business process management, and even the diagnosis of disease, all of which normally require human intelligence. Today, AI is integrated into our daily lives in several forms, such as personal assistants, automated mass transportation, aviation, computer gaming, facial recognition at passport control, voice recognition on virtual assistants, driverless cars, and companion robots.

AI provides tools creating intelligent machines which can behave like humans, think like humans, and make decisions like humans. The main goals of AI are [7]:
1. Replicate human intelligence
2. Solve knowledge-intensive tasks
3. Make an intelligent connection of perception and action
4. Build a machine which can perform tasks that requires human intelligence
5. Create some system which can exhibit intelligent behavior, learn new things by itself, demonstrate, explain, and can advise to its user.

The concept of AI is an umbrella term that encompasses many different technologies. AI is not a single technology but a collection of techniques that enables computer systems to perform tasks that would otherwise require human intelligence. AI has some branches, which are discussed next.

2.3 Components of AI

Each AI tool has its own advantages. Using a combination of these models, rather than a single model, is recommended. Figure 2.1 illustrates the AI tools. These tools are gaining momentum across every industry. There are plenty of businesses offering AI tools specifically designed for legal professionals.

1. *Expert Systems (ESs):* AI is sometimes a stand-alone independent electronic entity that functions much like human expert. ES was the first successful implementation of AI and may be regarded as a branch of AI mainly concerned with specialized knowledge intensive domain like medicine. An ES is computer software that simulates the judgment and behavior of a human expert. It is also known as intelligent system or knowledge-based system. It encapsulates specialist knowledge of a particular domain of expertise and can make intelligent decisions. It has a knowledge base and a set of rules that infer new facts from the knowledge. ESs solve problems with an inference engine that draws from a knowledge base equipped with information about a specialized domain, mainly in the form of if–then rules. It is based on expert knowledge in order to emulate human expertise in any specific field. The basic concept behind ES is that expertise (such as highly skilled medical doctor or lawyer) is transferred from a human expert to a computer system. Non-expert users, seeking advice in the field, question the system to get expert's knowledge [8, 9]. Figure 2.2 shows a typical ES.

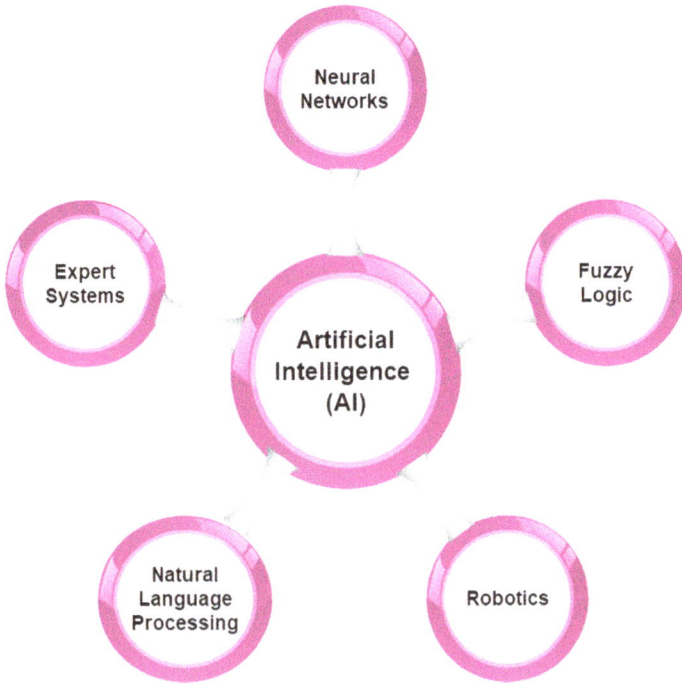

Figure 2.1: AI tools.

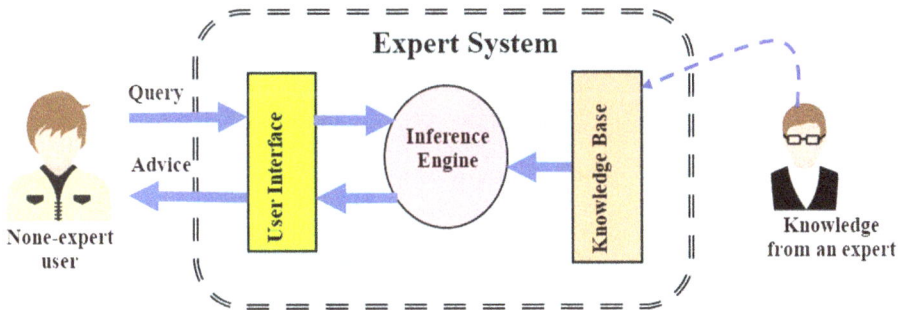

Figure 2.2: A typical expert system.

2. *Fuzzy Logic:* This makes it possible to create rules for how machines respond to inputs that account for a continuum of possible conditions, rather than straightforward binary. Where each variable is either true or false (yes or no), the system needs absolute answers. However, these are not always available. Fuzzy logic allows variables to have a "'truth value" between 0 and 1 so. It uses approximate human reasoning in knowledge-based systems. It was introduced in the 1960s by Lotfi Zadeh of University of California, Berkeley, known as the father of fuzzy set theory. Fuzzy logic is

useful in manufacturing processes as it can handle situations that cannot be adequately handled by traditional true/false logic [10].

3. *Neural Networks:* These are specific types of machine learning (ML) systems that consist of artificial synapses designed to imitate the structure and function of brains. An artificial neural network (ANN) is an information processing device that is inspired by the way the brain processes information. They were originally developed to mimic the learning process of human brain. The idea of ANNs was inspired by the structure of the human brain and by envy of what the brain can do. They may be regarded as a sort of parallel processor designed to imitate the way the brain accomplishes tasks. They are made up of artificial neurons, take in multiple inputs, and produce a single output. The network observes and learns as the synapses transmit data to one another, processing information as it passes through multiple layers [11]. As shown in Figure 2.3, ANNs are multi-layer fully-connected neural nets.

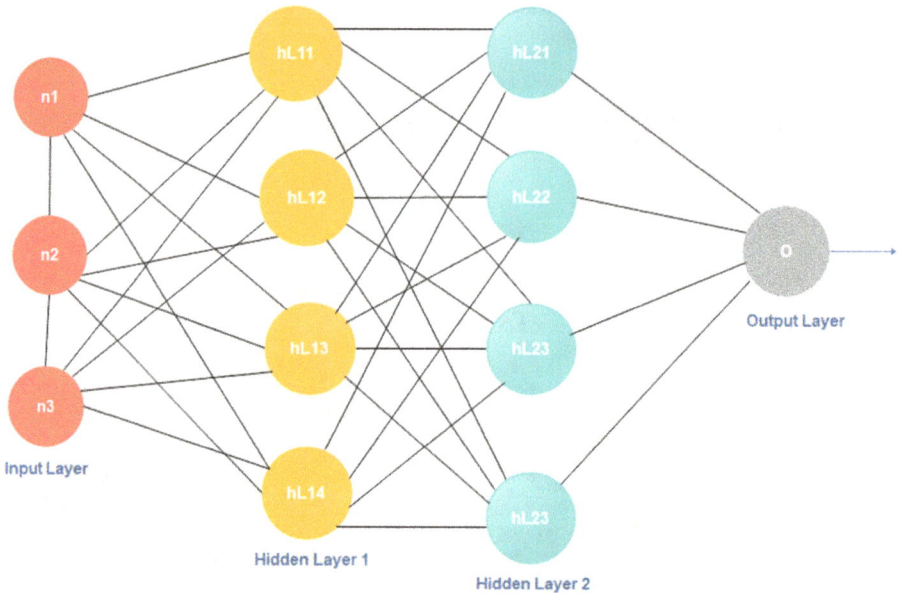

Figure 2.3: Artificial neural networks.

Different types of ANN are available: (1) support vector machine, (2) self-organization map, (3) multilayer perceptron. Typically, neurons are organized in layers. Due to the fact ANNs can reproduce and model nonlinear processes, they have found several applications in a wide range of disciplines including system identification and control, quantum chemistry, pattern recognition, medical diagnosis, finance, data mining, machine translation, neurology, and psychology [12].

4. *ML:* ML is essentially the study of computer algorithms that improve automatically through experience. It is the field that focuses on how computers learn from data. This includes a broad range of algorithms and statistical models that make it possible for systems to find patterns, draw inferences, and learn to perform tasks without specific instructions. ML is a process that involves the application of AI to automatically perform a specific task without explicitly programming it. Learning algorithms work on the assumption that strategies, algorithms, and inferences that worked well in the past are likely to work well in the future. ML techniques may result in data insights that increase production efficiency. Using ML can save time for practitioners and provide unbiased, repeatable results. Today, AI is narrow and mainly based on ML.

There are two types of learning: supervised learning and unsupervised learning. Supervised learning focuses on classification and prediction. It involves building a statistical model for predicting or estimating an outcome based on one or more inputs. It is often used to estimate risk. Supervised ML is where algorithms are given training data. Learning from data is used when there is no theoretical or prior knowledge solution, but data is available to construct an empirical solution. Supervised ML is increasingly being used in medicine such as in cardiac electrophysiology. Supervised learning includes both classification and numerical regression. In unsupervised learning, we are interested in finding naturally occurring patterns within the data. Unlike supervised learning, there is no predicted outcome. Unsupervised learning looks for internal structure in the data. Unsupervised learning algorithms are common in neural network models. ML techniques have been currently applied in the analysis of data in various fields including medicine, finance, business, education, advertising, cyber security, and energy applications [13, 14]. Figure 2.4 illustrates ML.

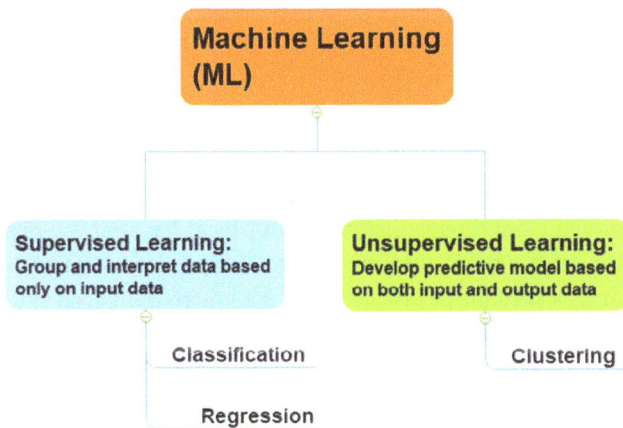

Figure 2.4: Machine learning.

5. *Deep Learning (DL):* The biggest breakthroughs for AI research have been in the field of ML, particular in the field of DL. This is a form of ML based on ANN. DL has enabled

many practical applications of ML. DL architectures are able to process hierarchies of increasingly abstract features, making them especially useful for purposes like speech and image recognition and NLP. DL networks can deal with complex non-linear problems. It extracts complex features from high-dimensional data and applies them to develop a model that relates inputs to outputs. The most common form of DL architectures is multi-layer neural networks. DL has many advantages over shallow learning. Due to this, DL networks have received much attention as they can deal with more complex non-linear problems.

Recently, companies such as IBM, Microsoft, Google, Apple, and Baidu have invested and developed DL. They have taken advantage of their massive data and large computational power to deploy DL on a large scale. Although DL has achieved some success and found applications in various fields, it is still in its infancy [15].

6. *Natural Language Processors:* For AI to be useful to us humans, it needs to be able to communicate with us in our language. Computer programs can translate or interpret language as it is spoken by normal people. Language is crucial around the world in communication, entertainment, media, culture, drama, movie, and economy. NLP refers to the field of study that focuses on the interactions between human language and computers. It is a computational approach to text analysis. It involves the study of mathematical and computational modeling of various aspects of language. It is an interdisciplinary field involving computer science, linguistics, logic, and psychology.

NLP is important because of the major role language such as English plays in human intelligence and because of the wealth of potential applications. NLP is commonly used for text mining, machine translation, and automated question-answering. Applications of NLP include interfaces to ESs and database query systems, machine translation, text generation, story understanding, automatic speech recognition, and computer-aided instruction. It also has great potential in healthcare, mobile technology, cloud computing, virtual reality, election, social work, and social networking [16].

7. *Robotics:* AI is heavily used in robots, which are computer-based programmable machines that have physical manipulators and sensors. Sensors can monitor temperature, humidity, pressure, time, record data, and make critical decisions in some cases. Robots have moved from science fiction to your local hospital. In jobs with repetitive and monotonous functions they might even completely replace humans. Robotics and autonomous systems are regarded as the fourth industrial revolution. In 1950, Isaac Asimov proposed the following three rules [17]:

1. A robot may not injure a human being, or, through inaction allow a human being to come to harm.
2. A robot must obey the orders given to it by human beings, except where such orders would conflict with the First Law.
3. A robot must protect its own existence as long as such protection does not conflict with the First or Second Law.

Robotics is a branch of engineering and computer science that involves the conception, design, manufacture, and operation of robots. A robot functions as an intelligent machine, meaning that, it can be programmed to take actions or make choices based on input from sensors. It involves using electronics, computer science, AI, mechatronics, and bioengineering. Robots are applied in many fields including agriculture, education, manufacturing, entertainment, medicine, industry, space exploration, undersea exploration, sex, power grid, agriculture, construction, meat processing, household, mining, aerospace, electronics, and automotive. For example, robot police with facial recognition technology have started to patrol the streets in China. Future robots will operate in highly networked environments where they will communicate with other systems such as industrial control systems and cloud services [18].

8. *Data Mining:* This deals with the discovery of hidden patterns and new knowledge from large databases. It may also be regarded as the process of discovering insightful and predictive models from massive data. It is an interdisciplinary subfield of computer science. Data mining exhibits a variety of algorithmic tools such as statistics, regression models, neural networks, fuzzy sets, and evolutionary models. Data mining has been applied with considerable success in business, retail industry, telecommunications, intrusion detection, biological data analysis, healthcare, geosciences, and computer security. Although it would not hurt to have some exposure to statistical analysis, one does not need to be an expert in statistics or a computer programmer to be a data miner [19].

9. *Computer Vision:* This is also known as machine vision. It is a scientific field which enables the machines to see. It processes images and makes computers understand contents in the image so that one can perform operations on the certain decisions. It helps transform otherwise ordinary factories into intelligent systems and formidable competitive weapons. It emphasizes the development of techniques which allow a computer to recognize or otherwise understand the content of a picture. It performs tasks such as object detection, recognition, tracking, facial recognition, etc. The goal of computer vision is to develop visual interpretive skills equivalent to those of humans. Although humans often take their visual skills for granted, visual interpretation requires intelligence. Computer vision made its way in all possible fields such as pattern recognition, ML, computer graphics, 3D reconstructions, virtual reality, and augmented reality. Computer vision provides the robot with its most flexible and powerful capability [20, 21]. Figure 2.5 shows computer vision and its subconcepts.

Using a combination of these models, rather than a single model, is recommended. Other areas of AI research in evolutionary computation, artificial general intelligence, and explainable AI. AI creation and implementation are transforming lives and cultures in Africa in a variety of ways including economically, socially, and politically. The application of AI in education, healthcare, agriculture, finance, and governance is showing a significant impact on the various sectors.

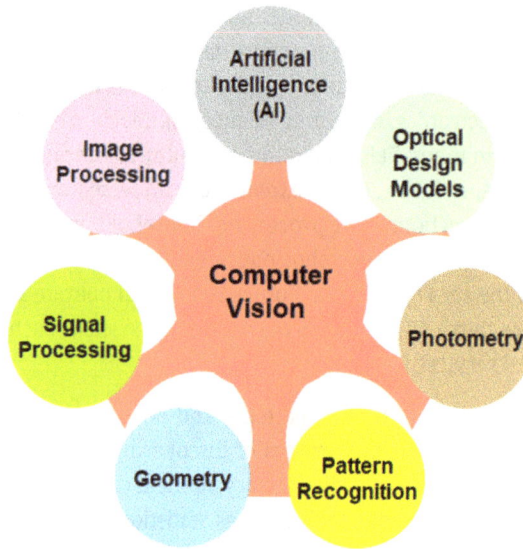

Figure 2.5: Computer vision and its subconcepts.

2.4 Use of AI in Africa

The AI revolution in Africa is not just a possibility; it is already underway. AI can be used in the production of pharmaceutical products, especially locally in African economies, to reduce the volume imported from other economies. For example, Chatbots in Kenya now provide healthcare services to people without visiting doctors. Mama Money and Mukuru enable easy and quick transfer of money across different countries in Africa. AI is a game-changing innovation with the potential to improve all sectors of the African social system. In Rwanda and Ghana, AI-powered diagnostic solutions are being deployed to improve medical imaging analysis.

However, the adoption and use of AI applications in African community raise some issues including skills acquisition, ethics, programming, data integration, user attitude, government policy, and insufficient infrastructure and network connectivity. Given the varied stages of AI adoption, implementation, and regulations across the continent's 54 nations, a cooperative and thoughtful approach is needed to standardize data and address these challenges effectively.

AI applications have not yet been widely adopted throughout Africa, with most African nations lacking the necessary elements required for technology adoption in the form of infrastructure. Generally speaking, Africa has been slow in the uptake of AI technologies, for a variety of reasons, from infrastructure challenges to limited financial resources. With only seven African nations (Benin, Egypt, Ghana, Mauritius, Rwanda, Senegal, and Tunisia) having drafted national AI strategies and none implementing formal AI regulation, substantial efforts are required to advance AI regulatory frameworks on the African continent. Unfortunately, harmful AI use has also been observed, for example, in Libya through the deployment of autonomous weapon systems and in Zimbabwe through facial recognition surveillance systems [22].

2.5 African Nations Adapting AI

There are several African countries that are beginning to have a dedicated strategy for AI. These countries are developing AI strategies and implementing data protection laws. Countries such as Egypt, Rwanda, and Mauritius have published comprehensive AI strategies. We examine the following selected African nations to see how they are adopting and implementing AI [23–25]:

- *Ghana:* Ghana is not doing badly on the Fourth Industrial Revolution. Indeed, Ghana has made significant strides towards developing AI technologies from a local context. Additional digital and AI innovations can be found in the health sector, for example services supplying medical supplies using drone technology. AI is also quite prevalent in agriculture, with tools developed to help farmers track weather patterns, monitor production conditions, and plan agricultural practices more efficiently. AI is also used to translate information into local languages. Ghanaian cashew farmers use drones to detect disease by collecting data from leaves. Before founding its AI lab in Ghana, for example, Google began working with farmers in rural Tanzania to understand some of the struggles they faced in maintaining consistent food production.
- *Mauritius:* Its AI strategy, published in 2018, describes AI and other emerging technologies as having the potential to address the country's social and financial issues. Areas of focus suggested in the strategy include manufacturing, healthcare, fintech, agriculture, and maritime traffic management.
- *Egypt:* Egypt published a comprehensive National Artificial Intelligence Strategy in July 2021, which is to be implemented over the next three to five years. Egypt has a national AI strategy (2021) built around a two-fold vision: exploiting AI technologies to support the achievement of Sustainable Development Goals (SDGs) and establishing Egypt both as a key factor in facilitating regional cooperation on AI and as an active international player. The strategy focuses on four pillars: AI for government, AI for development, capacity building, and international activities. Figure 2.6 shows Egypt National AI strategy [26].
- *Kenya:* Kenya's government started exploring the potential of AI in 2018 when it created the blockchain (or distributed ledgers technology) and AI task force (2019) to develop a roadmap for how the country can take full advantage of these technologies. As actions that could help achieve this goal, the report recommends investments in infrastructure and skills development and the development of effective regulations to balance citizen protection and private sector innovation. Egypt

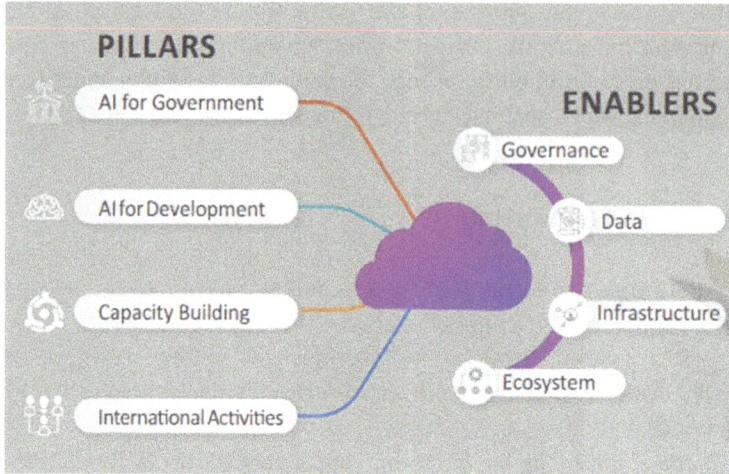

Figure 2.6: Egypt National AI strategy [26].

is documented to feature in the AI computing and research infrastructure dimension that has a total of 54 initiatives. The Kenyan Ministry of Information, Communication, and Technology (ICT), for example, has established an eleven-person "Blockchain and Artificial Intelligence Taskforce," to explore how the technologies can best be used to advance the country's development.

- *South Africa:* South African Artificial Intelligence Association (SAAIA) envisions a future where responsible and human-centric AI technologies are pivotal in South Africa's economic growth, trade, investment, equality, and inclusivity. Its advisory board members share a passion for responsible AI adoption and are active community builders addressing critical issues such as education, inclusion, training, regulation, ethics, policy, and investment. Individual membership to SAAIA is free, granting access to valuable resources, insights, and news throughout the year. An estimated more than 100 companies in South Africa are either integrating AI solutions into their existing operations or are developing new solutions.
- *Morocco:* The Policy Center for the New South is a Moroccan think tank aiming to contribute to the improvement of economic and social public policies that challenge Morocco and the rest of Africa. Through its analytical endeavors, the think tank aims to support the development of public policies in Africa and to give the floor to experts from the South. Only Tunisia and Morocco feature with each country having initiatives to provide funds for research.

2.6 Benefits

AI, which enables machines to exhibit human-like cognition, is unleashing the next wave of digital disruption. AI has the potential to fundamentally change the way businesses operate, drive innovation, and improve the lives of millions of people across Africa. The demography that has the longest experience with AI in Africa is the younger generation. AI technologies can be introduced into production value chains to increase productivity and strengthen production. AI-driven systems are also being deployed to fight organized crime.

In addition to the growing use of AI within surveillance systems across Africa, AI has also been integrated into weapon systems. Some African countries have used national security justifications to deploy surveillance technologies to control and discipline non-law abiding citizens. The application of AI in combat in northern Africa points to a future where AI-enabled weapons are increasingly deployed in armed conflicts in the region. Other benefits of AI in Africa include the following [27]:

1. *Sustainable Development:* In addition to enhancing economic growth, AI is playing an increasingly important role in sustainable development in Africa. The adoption of AI and related technologies in Africa could have the potential to significantly impact the achievement of the 2030 UN SDGs.
2. *Economic Growth:* The adoption of AI and related technologies in Africa could have the potential to significantly drive economic growth and improve access to quality education and healthcare, and promoting sustainable agriculture. AI can play a crucial role in addressing some of the continent's most pressing challenges.
3. *Mobil Tech:* Mobile technologies have permitted African nations to dramatically increase their communication capabilities while leapfrogging the need for old-fashioned infrastructure. The success of mobile technologies across Africa is prompting speculation among tech investors about whether AI applications will also take root in African nations. The near-ubiquity of mobile phones – nearly one per person – and the growing popularity of social media and messaging applications (apps) across Africa have made data more readily available. Mobile phone penetration in Kenya, for example, is 94%; almost a quarter of the country's households have an Internet connection – among the highest in the developing world. As shown in Figure 2.7, most African youths have a mobile phone [28].
4. *Digital Agenda:* AI-driven systems and solutions are commonplace across the continent. Countries from across different regions of Africa are increasingly developing or looking to develop national AI strategies to guide AI adoption. AI should build on national digital agendas and prioritize inclusive digital, data infrastructure and, skills development. Africa is playing a central role in the global AI supply chain, particularly in the early production phase. Countries, including Egypt, Rwanda, and Mauritius, have published comprehensive AI strategies. The implications of AI on gender equality must be carefully considered given the continent's digital agenda.

Figure 2.7: Most African youths have a mobile phone [28].

2.7 Challenges

While there are plenty of reasons to be optimistic about AI adoption and implementation in Africa, even its strongest proponents recognize that its introduction is likely to cause significant social and economic change. In spite of its vast potential, the adoption of AI in Africa face several challenges, including a lack of relevant technical skills, inadequate basic and digital infrastructure, insufficient investment in R&D, and a need for more flexible and dynamic regulatory systems. Education and skills training remain a critical challenge. Other challenges of AI in Africa include the following [29]:

1. *Daunting Challenge:* The daunting challenge of our times is to integrate AI into the lives of 1.4 billion people. This is an uphill task. In Africa, there are significant gaps in terms of access to knowledge, data, education, training, and human resources required for the development and the adoption of AI technologies primarily due to the digital divide. Only 8 out of the 54 countries in Africa have developed national AI strategies or policies. These countries are Rwanda, Benin, Egypt, Morocco, Mauritius, Tunisia, Sierra Leone, and Senegal.
2. *Low Connectivity:* At the forefront are related challenges such as difficult access to large amounts of data, quality of data, data storage, and data regulatory policy. AI technologies require a good quality connection and considerable energy requirements. Africa is one of the biggest places where we are not connected and when we are not connected, we cannot even talk about AI. For example, in Somalia, South Sudan, Ethiopia, and Congo, Internet penetration is below 20%. Failure to address the connectivity issue means that African countries will risk falling further behind at a critical time. Globally, Africa has the lowest average level of statistical capacity.

3. *Digital Literacy:* There is also the issue of digital literacy. The diffusion of AI across Africa is likely to be diverse and uneven. AI deployed in Africa relies on data that is not from the continent. AI experts have warned of a new "colonization" of the continent by this new technology if foreign companies continue to feed on African data without involving local actors. The critical factors necessary for AI technology to take hold are woefully absent across most of the continent, with possible exception Kenya, South Africa, Nigeria, Ghana, and Ethiopia.
4. *Fear:* AI critics often express fears that the technology could ultimately become uncontrollable and far surpass human capacities. There is a fear that the growth of AI and automation will lead to a displacement of jobs. Experts fear that Africa may end up with large multinationals in AI that will impose their solutions throughout the continent, leaving no room for creating local solutions [31].
5. *Lack of Infrastructure:* Many African nations still lack the statistical capacity, infrastructure, and good governance necessary to see AI take off. However, in a select handful of countries, AI solutions are already being successfully deployed at scale. What African governments need to do to strengthen the ecosystem necessary to see these technologies flourish?
6. *Lack of Reforms:* Many African countries remain incapable of requisite reforms in the areas of data collection and data privacy, infrastructure, education, and governance. Without those reforms, there is little chance that most African nations will be able to exploit AI.
7. *No Regulation:* There are insufficient regulations to protect against data misuse and to ensure personal privacy.
8. *Limited AI Expertise:* Africa lacks people who are trained in AI. It would be fair to say that AI expertise in Africa is rather limited, with few practitioners acquiring skills through formal education and training. Africa's underrepresentation in global AI provides an opportunity for Asian firms to tap business.
9. *Re-skilling:* The major policy response to the threat of job loss is to provide programs to re-skill or up-skill existing workforces.

2.8 Conclusion

Africa is a place where AI is used and developed: in social networks, in businesses, in healthcare, in agriculture, and in education. AI will revolutionize the way we do business across the African continent and transform life in Africa for the better. From agriculture to healthcare, AI is expected to enable faster and more profound progress in nearly every field of human endeavor and help address some of society's most daunting challenges. The knowledge of AI is still at an early stage and the population is still not certain of the advantages in developing African continent. Sound policy approaches are needed on AI implementation strategies to enable African nations build ecosystems that are inclusive, socially beneficial, and adequately integrated with on-the-ground realities [31, 32].

US technology multinationals are investing in AI and other emerging technologies because they recognize the potential impact these technologies can have on global health.

The involvement of external entities like IBM, Google, and Microsoft in promoting AI adoption in Africa is acknowledged. The uses of AI in Africa, and the potential benefits or harm, will depend on human users. Embracing the transformative power of AI is central to Africa's future. If African countries fail to find a way to harness AI's benefits, these economies could be left behind. African policy-makers should consider the best approach to AI strategies and adoptions in their countries. Africa is poised to lead and significantly influence the AI landscape. More information about AI in Africa can be found in the books in [33, 34].

References

[1] M. N. O. Sadiku, S. M. Musa, and U. C. Chukwu, *Artificial Intelligence in Education*. Bloomington, IN: iUniverse, 2022, p. ix.

[2] "Artificial intelligence for Africa," 2021, https://smartafrica.org/knowledge/artificial-intelligence-for-africa/

[3] "The top 5 emerging tech trends in Africa," https://www.linkedin.com/pulse/top-5-emerging-tech-trends-africa-nerdzfactoryorg

[4] S. Greenstein, "Preserving the rule of law in the era of artificial intelligence (AI)," *Artificial Intelligence and Law*, vol. 30, July 2021, pp. 291–323.

[5] M. N. O. Sadiku, "Artificial intelligence," *IEEE Potentials*, May 1989, pp. 35–39.

[6] "Artificial intelligence tutorial," https://www.javatpoint.com/artificial-intelligence-tutorial

[7] M. N. O. Sadiku, Y. Wang, S. Cui, S. M. Musa, "Expert systems: A primer," *International Journal of Advanced Research in Computer Science and Software Engineering*, vol. 8, no. 6, June 2018, pp. 59–62.

[8] "Expert system for medical diagnosis," https://tutorials.one/expert-systems-for-medical-diagnosis/

[9] G. Singh, A. Mishra, and D. Sagar, "An Overview of Artificial Intelligence," *SBIT Journal of Sciences and Technology*, vol. 2, no. 1, 2003.

[10] A. Dertat, "Applied deep learning – Part 1: Artificial neural networks," August 2017, https://towardsdatascience.com/applied-deep-learning-part-1-artificial-neural-networks-d7834f67a4f6

[11] Deloitte, "Artificial intelligence," March 2018, https://www2.deloitte.com/content/dam/Deloitte/nl/Documents/deloitte-analytics/deloitte-nl-data-analytics-artificial-intelligence-whitepaper-eng.pdf

[12] M. N. O. Sadiku, S. M. Musa, and O. S. Musa, "Machine learning, " *International Research Journal of Advanced Engineering and Science*, vol. 2, no. 4, 2017, pp. 79–81.

[13] "How machine learning algorithms make self-driving cars a reality," https://www.intellias.com/how-machine-learning-algorithms-make-self-driving-cars-a-reality/

[14] M. N. O. Sadiku, M. Tembely, and S. M. Musa, "Deep learning," *International Research Journal of Advanced Engineering and Science*, vol. 2, no. 1, 2017, pp. 77, 78.

[15] M. N. O. Sadiku, Y. Zhou, and S. M. Musa, "Natural Language Processing," *International Journal of Advances in Scientific Research and Engineering*, vol. 4, no. 5, May 2018, pp. 68–70.

[16] "Three laws of robotics," *Wikipedia*, the free encyclopedia https://en.wikipedia.org/wiki/Three_Laws_
      of_Robotics

[17] M. N. O. Sadiku, S. Alam, and S. M. Musa, "Intelligent Robotics and Applications," *International
      Journal of Trends in Research and Development*, vol. 5. No. 1, January-February 2018, pp. 101–103.

[18] M. N. O. Sadiku, A. E. Shadare, and S. M. Musa, "Data Mining: A Brief Introduction," *European
      Scientific Journal*, July 2015, vol. 11, no. 21, pp. 509–513.

[19] V. Kakani et al., "A Critical Review on Computer Vision and Artificial Intelligence in Food Industry,"
      Journal of Agriculture and Food Research, vol. 2, December 2020.

[20] https://www.researchgate.net/figure/AI-domains-see-online-version-for-colours_fig2_327215281

[21] D. Quinby, "Artificial intelligence and the future of travel," May 2017, https://www.phocus
      wright.com/Travel-Research/Research-Updates/2017/Artificial-Intelligence-and-the-Future-of-
      Travel

[22] "The state of AI in Africa report 2023," https://cipit.strathmore.edu/wp-content/uploads/2023/05/
      The-State-of-AI-in-Africa-Report-2023-min.pdf

[23] "Artificial intelligence in Africa: National strategies and initiatives," Unknown Source.

[24] "AI is here to stay! How artificial intelligence can contribute to economic growth in Africa,"
      https://reliefweb.int/report/world/ai-here-stay-how-artificial-intelligence-can-contribute-economic-
      growth-africa

[25] "South African Artificial Intelligence Association (SAAIA)," https://saaiassociation.co.za/

[26] "AI in Africa: Key concerns and policy considerations for the future of the continent," https://afripoli.
      org/ai-in-africa-key-concerns-and-policy-considerations-for-the-future-of-the-continent

[27] N. Allen and M. I. Okpali, "Artificial intelligence creeps on to the African battlefield," February 2022,
      https://brookings.edu/articles/artificial-intelligence-creeps-on-to-the-african-battlefield/

[28] Unknown Source.

[29] R. M. Mutiso, "AI in Africa: Basics over buzz," https://www.science.org/doi/10.1126/science.
      ado8276#:~:text=Lofty%20leapfrogging%20narratives%2C%20the%20idea,global%20trade)%2C%
      20with%20armies%20of

[30] "Interview: AI expert warns of 'digital colonization' in Africa," January 2024, https://news.un.org/en/
      story/2024/01/1144342#:~:text=UN%20News%3A%20Are%20there%20bad,room%20for%20creating
      %20local%20solutions.

[31] A. Gwagwa et al., "Artificial intelligence (AI) deployments in Africa: Benefits, challenges and policy
      dimensions," *The African Journal of Information and Communication*, vol. 26, 2020, pp. 1–28.

[32] A. Ade-Ibijola and C. Okonkwo, "Artificial intelligence in Africa: Emerging challenges," in D. O. Eke,
      K. Wakunuma, and S. Akintoye (eds.) *Responsible AI in Africa: Social and Cultural Studies of Robots and
      AI*. Palgrave Macmillan, 2023, pp. 101–117.

[33] D. O. Eke, K. Wakunuma, and S. Akintoye (eds.), *Responsible AI in Africa Challenges and Opportunities*.
      Palgrave Macmillan, 2023.

[34] S. Brokensha, E. Kotzé, and B. A. Senekal, *AI in and for Africa: A Humanistic Perspective*. Chapman &
      Hall/CRC, 2023.

Chapter 3
Robotics in Africa

Don't think of robots as replacements for humans – think of them as things that will help make us better at tackling many of the problems we face. – Eoin Treacy

3.1 Introduction

Africa is a continent that has 54 countries with an area of 30,370,000 km^2 and 1.4 billion individuals as of 2021, subdivided into 5 major regions, like Northern Africa (with countries like Libya, Egypt, North Sudan, Algeria, Morocco, and Tunisia as demonstrated), inhabiting the northerly region of Africa [1]. The continent is not just catching up with the world; it is propelling itself to the forefront of innovation. Africa is rising, and its tech scene is leading the way. Africa is closely watched as the next big growth market. It is the home to some of the youngest populations in the world.

Robotics constitutes one of the most exciting fields of technology today. It is the discipline of designing and constructing intelligent machines called robots. A robot is an autonomous mechanical device that is designed to sense its environment, carry out computations to make decisions, and perform actions like humans in the real world. Popular interest in robotics has increased in recent years. Robots are becoming more and more common in our society and more integrated into our lives. This is due to the fact that they are becoming smarter, smaller, cheaper, faster, more flexible, and more autonomous than ever before. The robotic revolution is going to change us as humans [2, 3]. In recent years, the robotics and artificial intelligence (AI) fields have grown and changed uniquely and impacted many industries and sectors worldwide. Although Africa is quickly adopting new technologies, significant problems still make it hard for robots and AI to be widely used and integrated.

Robots are no longer something you see in the movies and TV shows. They are currently in homes and businesses. They are becoming more and more common in our society and more integrated into our lives. Some pioneering innovators in Africa are making significant strides in the development of robotic and mechanical systems designed to revolutionize the harvesting of diverse crops. African Union (AU) Member States are encouraged to invest in R&D to enhance the affordability and accessibility of these robots for African farmers. Several African nations have been investing in building their own capabilities in robotics R&D. Universities and research centers are conducting groundbreaking research in areas like autonomous systems, AI, and human-robot interaction [4].

This chapter examines the adoption of robotics in African nations. It begins with describing what robots are and presents different types of robots. It presents the uses of robots in Africa. It discusses how robots are adapted in some African nations. It highlights the benefits and challenges of robots in Africa. The last section concludes with comments.

https://doi.org/10.1515/9783112211984-003

3.2 What Is Robotics?

Robotics is a relatively new field that is dedicated to the design, construction, and use of robots. It is a technology field that uses electronic or mechanical technology to replace human labor. Robots are machines with enhanced sensing, control, and intelligence used to automate, augment, or assist human activities. They are currently used in manufacturing and production firms. They are expanding to other business industries.

The word "robot" was coined by Czech writer Karel Čapek in his play in 1920. Isaac Asimov coined the term "robotics" in 1942 and came up with three rules to guide the behavior of robots [5]:

1. Robots must never harm human beings.
2. Robots must follow instructions from humans without violating rule 1.
3. Robots must protect themselves without violating the other rules.

Robots are becoming increasingly prevalent in almost every industry, from healthcare to manufacturing. They are regarded as intelligent agents that can perform actions similar to what humans can do. Robotics technology has been implemented in a variety of fields including manufacturing, medicine, elderly care, rehabilitation, education, agriculture, home appliances, search and rescue, car industry, defense, and more.

Robotics is an interdisciplinary discipline embracing mechanical engineering, electrical engineering, computer science, and others. The goal of robotics is to create intelligent machines (called robots) that behave and think like humans. Robots were originally intended for use in industrial environments to replace humans in tedious and repetitive tasks. They are regarded as intelligent agents that can perform actions like what humans can do. Today, robots help human beings in everyday life. Depending on applications, there are many types of robots. Robotics technology has been implemented in a variety of fields, including manufacturing, medicine, elderly care, rehabilitation, education, agriculture, home appliances, search and rescue, car industry, and defense.

3.3 Types of Robots

Robotics has advanced and taken many forms, including fixed robots, collaborative robots, mobile robots, industrial robots, medical robots, police robots, military robots, officer robots, service robots, space robots, social robots, personal robots, and rehabilitation robots. The development of robotics has been driven by the need for automated systems that can perform tasks without human intervention. This has led to the creation of a wide variety of robots. Thus, robotics can take on a number of forms, leading to many types of robots. They are designed for different environments and various applications. The commonly used types of robots are as follows [6–9]:

1. *Humanoid Robots*: These are robots that mimic human behavior. They usually perform human-like activities (like walking, running, jumping, and carrying objects).

2. *Autonomous Robots:* These operate independently of human operators. They require no human supervision. An example of an autonomous robot is the vacuum cleaner, shown in Figure 3.1 [10].

Figure 3.1: An autonomous robot is a vacuum cleaner [10].

3. *Teleoperated Robots:* These are semiautonomous bots that allow human control from a safe distance using wireless networks.

4. *Educational Robots:* These are used to teach programming and engineering principles to students. They encompass the entire span of K-12 grades. Robotics helps kids learn to think creatively and critically. It also helps them increase their problem-solving abilities. Examples of educational robots are shown in Figure 3.2 [11].

Figure 3.2: Examples of educational robots [11].

5. *Agricultural Robots:* Agricultural robots are used to automate tasks in agriculture. The use of robots in agriculture is closely linked to the concept of AI-assisted precision agriculture and drone usage. They are used for tasks such as planting, harvesting, and monitoring crops.

6. *Collaborative Robots:* These robots (known as cobots) are flexible and easily reprogrammable on the fly. They can learn complex tasks and then act as a collaborator with the skilled workers. They are capable of avoiding unwanted collisions and they can recognize when they have bumped into something. Collaborative robots may be regarded as the friendly face of workplace automation.

7. *Drones:* DRONE (Dynamic Remotely Operated Navigation Equipment) is commonly referred to as unmanned aerial vehicle (UAV). Drones are equipped with all the software, sensors, and hardware that a farmer will need to check the health of crop and survey farmland. A drone typically consists of propulsion and navigation systems, GPS, sensors, infrared cameras, software, and programmable controllers. Drones are flying robots, a type of robots, that are poised to proliferate in certain commercial sectors. Drones can help utility crews after a storm by quickly and safely identifying areas in need of repair. Drones can also help with maintenance tasks, such as surveying solar panels for damage.

8. *Chatbots:* These robots carry out simple conversations such as in a customer service setting. Chatbots have empowered the banks and other financial institutions by simplifying the complex processes. We interact with Facebook Messenger bots all the time. Messenger bots are revolutionizing the small business world. Messenger bots can answer customers' questions, collect user's information, organize meetings, reduce overhead costs, and engage in other business tasks. Big companies such as Walmart, Alibaba, and Amazon have been benefitting the help of bots.

9. *Industrial Robots*: The most common use of robots in industry is for simple and repetitive tasks. The role of robots is becoming substantial for industrial applications. Industrial robots are used in manufacturing, assembly, and other industrial processes. Examples of industrial robots include assembly line processes, picking and packing, welding, and similar functions. Robots can reduce the risk of injury to humans in dangerous work environments. Examples of industrial robots are displayed in Figure 3.3 [12].

10. *Military Robots:* In the military sectors, robotic technology is being applied in many areas. More recent developments mean that military forces worldwide use robots in areas such as UAVs, unmanned ground vehicle, drones, and surveillance. Military drones are flying over areas of war and conflict, in hostage situations, and for natural and manmade disasters. The military also employs robots to (1) locate and destroy mines on land and in water, (2) enter enemy bases to gather information, and (3) spy on enemy troops.

Figure 3.3: Examples of industrial robots [12].

11. *Exploration Robots:* Robots are often used to reach hostile or inaccessible areas. A good example of exploratory robots is in space exploration. Robots can go to the planets. They can be used to explore space.

12. *Entertainment Robots:* Robots can be used in entertaining audiences. Increasingly (particularly during the pandemic), people are buying robots for enjoyment. There are several popular toy robots, and there are even robot restaurants and giant robot statues.

13. *Hybrids:* The various sorts of robots are often combined to create half-breed solutions capable of additional intricate tasks.

Other types include manipulators, medical robots, nanorobots, construction robots, swarm robots, domestic robots, mobile robots, fixed robots, service robots, social robots, rehabilitation robots, underwater robots, field robots, self-driving vehicles, cloud robots, police robots, officer robots, space robots, personal robots, and sailboat robots [13].

3.4 Use of Robotics in Africa

In recent years, several organizations in Africa have launched initiatives to advance participation in robotics. In some parts of Africa, robots are used in agricultural harvesting, mining, controlling traffic, and even fighting deadly diseases. As technology and robotic automation spread across Africa, young children have been attracted to creative designs. These students have developed remarkable robots that address real-world challenges. They have learned that building a robot is a team effort, as displayed in Figure 3.4 [14]. They have been meticulously designed to solve daily problems faced by individuals, showcasing how technological advancements can enhance our lives.

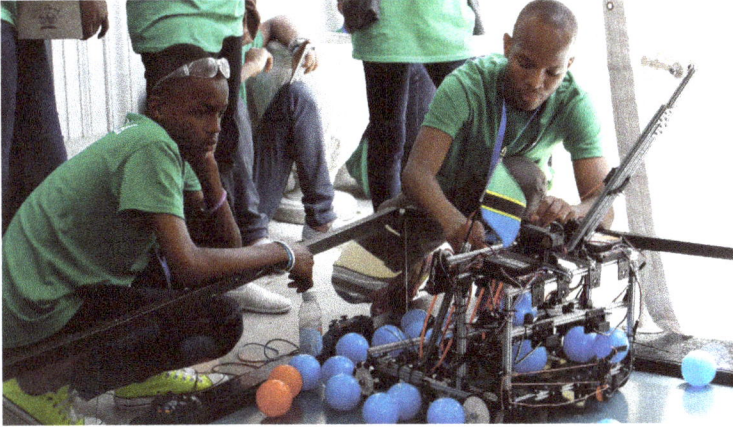

Figure 3.4: Building a robot is a team effort [14].

Carnegie Mellon University-Africa aims to use AI and robotic systems to address important problems in areas such as transportation, building systems, manufacturing, energy, agriculture, security, health, and climate. (It is believed that Pittsburgh, Pennsylvania, is the birthplace of AI and a global leader in the growing robotics industry, with Carnegie Mellon University as a trailblazer since the 1950s.) The goal is to improve our understanding of real-world systems and address specific challenges that can have a positive impact on both society and the environment. A culturally sensitive social robot for Africa is shown in Figure 3.5 [15].

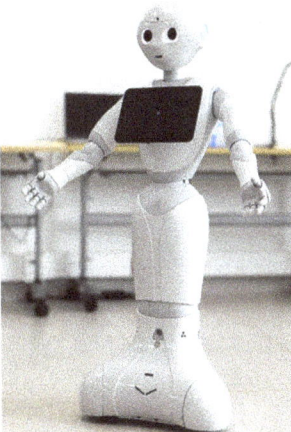

Figure 3.5: A culturally sensitive social robot for Africa [15].

The African Robotics Network (AFRON) is a community of institutions, organizations, and individuals engaged in robotics in Africa. It was established in April 2012 to promote communication and collaborations that will enhance robotics-related education, research, and industry on the continent. In order to achieve this, AFRON organizes projects, meetings, and events in Africa and at robotics and automation conferences abroad.

Pan-African Robotics Competition (PARC) is an annual youth robotics competition that brings together middle school, high school, university, and young professional robotics team across Africa and its diaspora. PARC recently launched a virtual learning platform allowing youth around the world to engage with coding, programming, and robot design with online and offline capabilities.

Collaborative robots are designed to work seamlessly with workers in all fields; they facilitate improvements in workers' efficiency. Cobots are being integrated into a wide range of disruptive digital manufacturing innovations that allow a factory to perform smarter and produce better quality material. Cobots can also provide logistic solutions, and their possibilities for manufacturing are limitless. Figure 3.6 shows a typical cobot [16]. Besides manufacturing, a great number of other industries across Africa have already adopted collaborative robots.

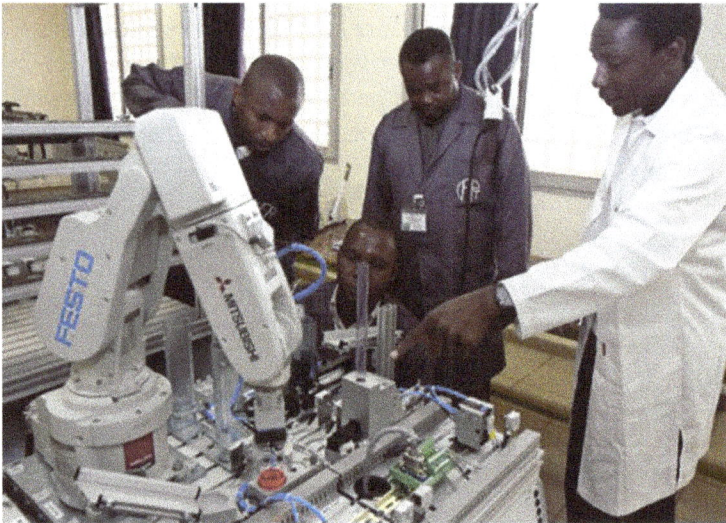

Figure 3.6: A typical cobot [16].

3.5 Adapting Robotics in African Nations

There are several African countries that are beginning to have a dedicated strategy for adopting robots. We present the following African nations as typical examples [17]:

– *Nigeria:* In Nigeria, the National Center for Artificial Intelligence and Robotics has significantly pushed the country to propel in machine learning, the Internet of things, blockchain technology, intelligent robotics, and digital manufacturing and prototyping. Uniccon Group unveiled Africa's first humanoid robot in Abuja, Nigeria. Omeife, the 6-foot-tall female battery-powered human-like robot, is African by design and has Igbo-like physical attributes. It can speak different languages like Igbo, Yoruba, English, French, Swahili, Wazobia, Pidgin, Afrikaans, and Arabic [18]. In Nigeria, there is a growing interest in agricultural robotics, with large-scale farms utilizing harvesting robots to boost productivity. The National Center for Artificial Intelligence and Robotics has significantly pushed the country to propel in machine learning, the Internet of things, blockchain technology, intelligent robotics, extended reality, and digital manufacturing and prototyping. Omeife is the world's first African humanoid robot, as shown in Figure 3.7 [19]. A team of engineers and programmers created her at the University of Nigeria, Nsukka.

Figure 3.7: Omeife is the world's first African humanoid robot [19].

– *Ghana:* The Kwame Nkrumah University of Science and Technology (KNUST) has introduced the TEK mechanical harvester. This groundbreaking technology has the remarkable capability to harvest one cassava plant in just 1 s, with the potential to significantly streamline the cassava harvesting process across the continent.
– *Uganda:* In early 2016, a pair of sprawling posture robots were designed, one designed to mimic a crocodile and another designed to mimic a monitor lizard, along the banks of the Nile River in Uganda, Africa. These robots fell at the intersection of our interests in developing robots to study animals and robots for disaster response. The robots needed to be designed on the basis of a systematic study of data on the model specimens, be fabricated rapidly, and be reliable and robust enough to handle what the wild would throw at them [20].

- *South Africa:* Robots are used in the gold mining to eliminate the associated risk involved in these jobs. Robots now replace humans to assess the depth of some of the country's gold mines. South African engineers at the Council for Scientific and Industrial Research (CSIR) in Pretoria are currently testing robots that will be able to assess the safety of mines after they have been blasted. Spot is a mobile ground robotics tool that is helpful in the South African market in that their legacy mining methodologies do not easily facilitate the use of autonomous drones.
- *Congo:* The Democratic Republic of Congo has introduced robotic machines into the public sector. In the capital city of Kinshasa, authorities have installed two 8-foot-tall solar-powered robots ("robocops") to help direct traffic and prevent road accidents. These robots have eliminated the need for human traffic wardens as they can detect pedestrians and are designed to withstand all weather conditions.
- *Botswana:* An African nation, known for gorgeous diamonds, has introduced robots into its mining sector. These robots are designed to perform tasks and go into depths that human miners cannot simply reach, bringing up stones.

3.6 Benefits

By embracing robotics, African nations have the potential of addressing critical issues of poor harvesting practices and food wastage, bolstering agricultural efficiency. This will ultimately contribute to ensuring a sustainable and abundant food supply for its growing population. Having a national policy guiding AI and robotics adoptions is essential. Robots do the heavy lifting while humans leave or make their way to the working face after or before a shift. Robots are the ideal solution for "no-go areas." The overriding benefits are those that involve efficiencies and precision while achieving zero harm. In Africa, robotics is being applied in various sectors such as agriculture, healthcare, education, manufacturing, and logistics. For example, robotic devices can assist healthcare workers in performing surgeries or providing care to patients in remote areas. Other benefits of robotics in Africa include the following [21]:

1. *Cost:* On a continent where money for expensive robots is limited, some think that there is real potential for Africa to build a reputation for coming up with more affordable, accessible robots. Low-cost educational robots have made robotics more accessible to young people.

2. *Manufacturing:* Africa's journey toward becoming a hub for car manufacturing requires a paradigm shift, and robotics can be the catalyst for this transformation. The continent may not be maximizing its labor force to do the jobs currently being taken over by robots. Instead, everything African workers could have done, robots can do better and faster. Industrial robots and AI are increasingly threatening manufacturing in emerging markets. At the moment, Africa only has a regional average of 2 indus-

trial robots per 100,000 manufacturing workers. As China did at the end of the last century, Africa could have taken advantage of relatively cheap, semiskilled labor in its youthful population, finally diversifying the continent's economies into manufacturing and services as engines for growth. The emergence of green technology has once again put African manufacturing in the spotlight. The shift of manufacturing from China to Africa is happening even without the prompting of the Chinese government. This is due to the rapid decline of China's population. Projections indicate that the population will continue its dramatic decline, going from the current 1.4 billion to 1 billion in 2080, and 800 million in 2100.

3. *Automation:* The robotic automation of manufacturing will also give rise to new job opportunities in the design, construction, vending, installation, management, and continued maintenance of robots. Automation is technology that assists humans, with limited guidance, in the production, maintenance, or delivery of products or services. An era characterized by a rise of autonomous robots and self-learning software is upon us. Many industrialized economies are being transformed by the increasing automation of work. Self-driving cars upending the taxi and trucking industries will be one of the most visible signs of these changes in the near future. The potential for job displacements due to automation is an important concern. Yet, sub-Saharan Africa does have areas of economic activity, where the economic calculus favors automation.

4. *Education:* A side benefit of the spread of robots is its use in education. There are initiatives aimed at promoting robotics education and entrepreneurship across Africa. Many institutions ranging from local high schools to universities around the continent are integrating programs like robotics and AI into their curriculum. For example, RoboCupJunior is an educational robotics initiative that aims to enhance learning through educational robotics activities around the world. RoboCupJunior has three distinguished leagues – Soccer, Rescue, and Dance – that attract students from all the continents. Robotics can also be used as a tool for teaching STEM (science, technology, engineering, and mathematics) subjects, helping to inspire and educate the next generation of African innovators. However, Africa has not taken significant advantage of this initiative with a rather low participation of a few African countries.

5. *Agriculture:* Agriculture is one of the key areas where robotics is making an impact in Africa. The agricultural sector stands as a cornerstone of African economies, serving as a multifaceted generator of income through foreign trade, employment opportunities, and sustenance. Since a significant portion of Africa's population relies on agriculture for their livelihoods, there is great potential for robotics to improve efficiency and productivity in farming practices. In spite of the enormous potential of Africa as a breadbasket, the agricultural sector of the continent is faced with constraints that hinder its progress and the welfare of its populations. African countries are confronted with alarming levels of food loss. The quest for a promising and trans-

formative frontier in African agriculture could be enhanced by the adoption and integration of robotics in agriculture through modern farming practices such as mechanization. Robotics-based mechanization holds significant promise, especially in the realm of harvesting, as shown in Figure 3.8 [22]. African farmers can use robotic harvesting to enhance food security in the region. Several AU Member States are adopting agricultural robotics to modernize their farming practices.

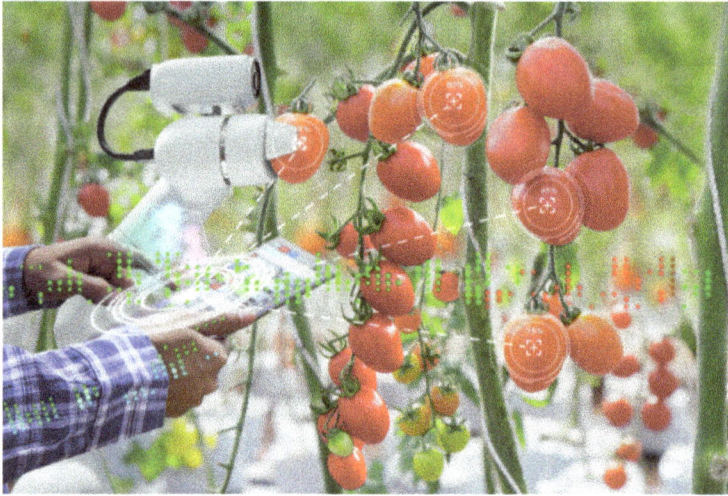

Figure 3.8: Robotics-based mechanization for harvesting [22].

6. *Globalization:* The world is becoming one. Globalization is slowly causing the interaction and integration of companies, governments, and people of different nations. In this process, technology, products, and information are all spread at a faster pace. There are new developments that are transforming the way people live, work, and relate to one another in South Africa. This is shaped by the disruptive technologies such as AI and robotics [23].

7. *Assistance:* Assistive robots are transforming human lives in multiple ways, according to the UN agency. Using machine learning and AI, these robots offer support in mobility, communication, self-care, and other essential daily tasks, giving people who need it renewed confidence and autonomy. Robots like Grace can support people's health and well-being, provide high-quality educational services, reduce inequalities by helping persons with disabilities, reduce waste, help build resilient infrastructure, and broadly enhance social good, according to the UN telecommunications agency. The journey toward a technologically advanced Africa begins with the realization that robotics is not just a tool; it is the key to unlocking the continent's boundless potential.

8. *Job Creation:* Embracing robotics industries not only enhances productivity but also creates opportunities for skill development and employment. As African nations invest in the deployment and maintenance of robotic systems, there is a growing demand for skilled technicians, engineers, and innovators.

3.7 Challenges

There are certainly challenges to overcome on using robots in Africa, such as limited access to resources and infrastructure, and the growing interest and investment in robotics in Africa bode well for the continent's future development. African countries are importing robots and people feel that their jobs are at risk. With continued support and collaboration, Africa has the potential to become a significant player in the global robotics industry. The AU High-Level Panel on Emerging Technologies (APET) recommends that AU Member States develop supportive and enabling regulations aimed at reducing bureaucratic burdens and providing tax incentives to farmers who invest in robotics technology. The challenges of embracing robotics industries in Africa are not to be underestimated. Some of those challenges include the following [24]:

1. *Food Security:* Food is a fundamental need every person must have in order to survive and participate actively in society. Food security occurs when all people in a community have access to sufficient, safe, and affordable food. Food-secure individuals and households do not have to resort to begging, stealing, or scavenging for food [25]. African nations face significant challenges in accomplishing food security due to factors such as a growing population, urbanization, and limited agricultural and food production capacity. In terms of promoting food security, Google AI lab has collaborated with farmers in rural Tanzania to create a machine learning application called "Nuru" (meaning "light" in Swahili) to diagnose early stages of cassava plant diseases for the advancement of the production of a common staple crop that provides food for over 500 million people.

2. *Fear:* Robots are set to take Africa's manufacturing jobs even before it has enough. Fear of losing jobs to computers is common, but robotics can allow many jobs to evolve besides ensuring safer, efficient, and better work results. Robots may take away jobs, but they could also create new ones. In fact, robots will create more jobs that did not exist before. Robotics developed in Africa could help local economies.

3. *Infrastructure:* Infrastructure, funding, and skill gaps need to be addressed strategically. Public-private partnerships, government incentives, and collaborations with international organizations can pave the way for overcoming these barriers and unlocking the full potential of robotics in African industries.

4. *Environmental Sustainability:* Robotics also align with the growing emphasis on environmental sustainability. Automated systems can be designed to optimize energy usage and minimize waste, contributing to eco-friendly and sustainable industrial practices. This not only meets global environmental standards but also positions African industries as responsible contributors to the global community.

5. *Structural Inequalities:* These are the disproportionate levels of access to socioeconomic and political resources such as education, employment, income, information and communications technologies, and healthcare. When it comes to access to such socioeconomic resources, African countries are among the least developed in the world. Digital divides in Africa are linked to issues such as inadequate telecommunications network, lack of supporting infrastructure like electricity, unaffordability of smartphones, and lack of digital skills.

3.8 Conclusion

Robotics is a branch of engineering that involves the conception, design, manufacture, and operation of robots. Robotics can take on a number of forms, with the end objective being that of assisting humans with intelligent machines. Today, there are robots that can autonomously sense, reason, plan, act, move, communicate, and collaborate with other robots. The robotic revolution is going to change us as humans. Robotics in Africa is an exciting and rapidly growing field with a lot of potential for innovation and development. While Africa may not be as commonly associated with robotics as some other regions of the world, there are numerous initiatives, projects, and startups emerging across the continent that are driving progress in the region.

Africa is getting ready for the impending age of robots. It is emerging as a key player in the robotics market, with countries like South Africa leading the way in developing innovative robotic solutions for various industries. For more information about robotics in Africa, one should consult the books in [26–28] and the following related journals devoted to robotics:

- *Robotica*
- *Robotics and Autonomous Systems*
- *Robotics and Computer-Integrated Manufacturing*
- *Advanced Robotics*
- *Autonomous Robots*
- *Journal of Robotics*
- *Journal of Robotic Systems*
- *Journal of Robotic Surgery*
- *Journal of Robotics and Mechatronics*
- *Journal of Intelligent & Robotic Systems*
- *Journal of Mechanisms and Robotics-Transactions of the ASME*
- *Journal of Automation, Mobile Robotics and Intelligent Systems*

- *Journal of Future Robot Life*
- *IEEE Robotics and Automation Letters*
- *IEEE Transactions on Robotics*
- *International Journal of Medical Robotics and Computer Assisted Surgery*
- *International Journal of Robotics Research*
- *International Journal of Social Robotics*
- *International Journal of Humanoid Robotics*
- *International Journal of Advanced Robotic Systems*

References

[1]   W. Shafik, "Chapter 7 Navigating Emerging Challenges in Robotics and Artificial Intelligence in Africa," https://www.irma-international.org/viewtitle/339985/?isxn=9781668499627
[2]   M. Thomas, "The future of robots and robotics," February 2021, https://builtin.com/robotics/future-robots-robotics
[3]   M. N. O. Sadiku, S. Alam, and S. M. Musa, "Intelligent robotics and applications," *International Journal of Trends in Research and Development*, vol. 5. No. 1, January-February 2018, pp. 101–103.
[4]   M. N. O. Sadiku, U. C. Chukwu, and J. O. Sadiku, "Robotics in Africa," *Innovative: International Multidisciplinary Journal of Applied Technology*, vol. 2, no. 4, 2024, pp. 88–95.
[5]   "Human–robot interaction," *Wikipedia*, the free encyclopedia https://en.wikipedia.org/wiki/Human–robot_interaction
[6]   "Robotics," *Wikipedia*, the free encyclopedia https://en.wikipedia.org/wiki/Robotics
[7]   R. D. Davenport, "Robotics," in W. C. Mann (ed.), *Smart Technology for Aging, Disability, and Independence*. John Wiley & Sons, 2005, Chapter 3, pp. 67–109.
[8]   M. N. O. Sadiku, S. Alam, and S. M. Musa, "Intelligent robotics and applications," *International Journal of Trends in Research and Development*, vol. 5, no. 1, January-February 2018, pp. 101–103.
[9]   A. Gautam, "What is robotics? Its types and applications," January 2023, https://www.electronics foru.com/tech-zone/tech-of-robotics/robotics-types-applications
[10] S. Daley, "Robotics technology," August 2022, https://builtin.com/robotics
[11] "8 Educational robotics kits we'll always recommend," https://www.eduporium.com/blog/8-robotics-brands-well-always-recommend-for-education/
[12] Robotics Tomorrow, "Industrial robots push into new applications and industries," August 2022, https://www.automationalley.com/articles/industrial-robots-push-into-new-applications-and-industries
[13] T. R. Kurfess (ed.), *Robotics and Automation Handbook*. Boca Raton, FL: CRC Press, 2005.
[14] L. Chutel, "Robots are set to take Africa's manufacturing jobs even before it has enough," July 2017, https://qz.com/africa/1037225/robots-are-set-to-take-africas-manufacturing-jobs-even-before-it-has-enough
[15] P. McSharry, C. Tucker, and D. Vernon, "Carnegie Mellon University Africa," https://www.africa.engi neering.cmu.edu/research/artificial-intelligence.html
[16] "African robotics revolution: How robots will help Africa," https://www.mobilevillage.com/african-robotics-revolution/
[17] T. Idowu, "African countries are importing robots and young people's jobs are at risk," April 2018, https://www.cnn.com/2017/08/22/africa/robots-in-africa/index.html
[18] "Africa is getting ready for the impending age of robots," December 2022, https://techcabal.com/2022/12/12/africa-robots/

[19] B. Johnson, "Meet Omeife: The World's First African Humanoid Robot," January 2023, https://www.funtimesmagazine.com/2023/01/24/424444/meet-omeife-the-worlds-first-african-humanoid-robot

[20] K. Melo, T. Horvat, and A. J. Ijspeert, "Animal robots in the African wilderness: Lessons learned and outlook for field robotics," *Science Robots*, vol. 8, no. 85, December 2023. https://www.science.org/doi/10.1126/scirobotics.add8662

[21] S. Macnamara, "Robotics – a slow adoption, and yet so many benefits!" https://www.africanmining.co.za/2022/05/03/robotics-a-slow-adoption-and-yet-so-many-benefits/

[22] "Transforming African agriculture through adoption of robotics technology," October 2023, https://www.nepad.org/blog/transforming-african-agriculture-through-adoption-of-robotics-technology

[23] M. B. Rapanyane and F. R. Sethole, "The rise of artificial intelligence and robots in the 4th Industrial Revolution: implications for future South African job creation," *Contemporary Social Science*, vol. 15, no. 4, August 2020, pp. 489–501.

[24] E. O. Arakpogun et al., Artificial intelligence in Africa: Challenges and opportunities," https://researchportal.northumbria.ac.uk/ws/portalfiles/portal/31309999/AI_in_Africa_Opportunities_and_Challenges_Paper_68_Manuscript.pdf

[25] M. N. O. Sadiku, S. M. Musa, and O. S. Musa, "Food security: A primer," *Invention Journal of Research Technology in Engineering and Management*, vol. 2, no. 7, July 2018, pp. 16–19.

[26] R. Mwanaka (ed.), *Writing Robotics: Africa Vs Asia Vol 2*. Mwanaka Media and Publishing Pvt Limited, 2020.

[27] *2021 Rapid Product Development Association of South Africa Robotics and Mechatronics Pattern Recognition Association of South Africa (RAPDASA RobMech PRASA)*. IEEE Press, 2021.

[28] M. N. O. Sadiku, *Robotics and Its Applications*. Moldova, Europe: Lambert Academic Publishing, 2023.

Chapter 4
Drones in Africa

Drones overall will be more impactful than I think people recognize, in positive ways to help society. – Bill Gates

4.1 Introduction

Africa is one of the most complex regions in the world. Yet, its potential for future prosperity is nothing short of astounding. Although it has access to abundant natural resources, including diamonds, gold, silver, copper, iron, oil, salt, sugar, cocoa beans, wood, and tropical fruits, Africa paradoxically remains one of the poorest and least-developed continents in the world. This is the result of a range of factors, such as poor central planning, governmental corruption, lack of infrastructure, low levels of public education, little foreign investment, and frequent conflicts between tribes [1].

Due to its size and proximity to the equator, the climate of Africa ranges from the arid deserts in the north to dense rain forests in the south. Sub-Saharan Africa is perceived to be one of the world's most deprived regions with a history of civil unrest, corrupt governance, low educational enrolment levels, poor healthcare delivery, a wide digital divide, as well as lacking infrastructure to meet present-day global socioeconomic demands [2]. Africa is rising, and its tech scene is leading the way. It is closely watched as the next big growth market. It is the home to some of the youngest populations in the world.

Drones are autonomous or remotely controlled multipurpose aerial vehicles driven by aerodynamic forces. They are devices that are capable of sustained flight and do not need a human onboard. Typically, a drone consists of an air frame, propulsion system, communication system, and navigation system [3]. Common drone configurations include fixed-wing, rotary-wing, multirotor, and hybrid designs. Their obvious usage areas are transportation and package delivery. As illustrated in Figure 4.1, compared with other transport modes, drones are the fastest and the least expensive [4]. Since a drone can fly over an inaccessible road, organizations have begun to use drones for healthcare delivery. While governments and regulators may be cautious about allowing drones to roam the skies, healthcare deliveries have a compelling reason to go for it.

Drones have the potential to play a transformative role globally. In recent times, the drone technology has gradually gained widespread adoption all over the world, including Africa. It has found diverse applications across various industries. Drones were used by the United States to track down and kill Islamic militants in Somalia. The United Nations is using drones to support military peacekeeping operations and security services in the Democratic Republic of Congo, Mali, and the Central African

https://doi.org/10.1515/9783112211984-004

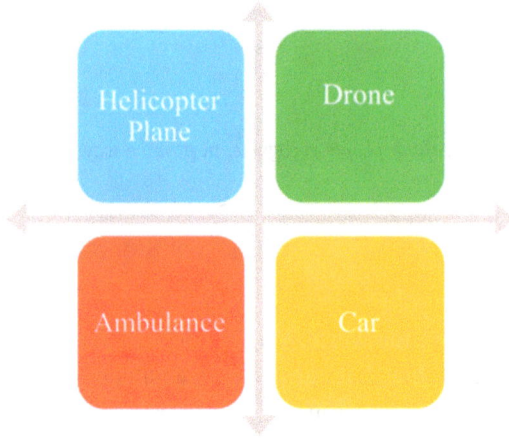

Figure 4.1: Drone is compared with other means of transportation [4].

Republic of Congo [5]. Amazon announced its plan to use drones to deliver packages to customers.

Drones offer huge potential for Africa in terms of economic development and human security, such as monitoring weather patterns. The usage of drone technology is gaining attention in Africa. Africans are employing drones for commercial, humanitarian, and military purposes. Drone technology is providing the delivery solutions that can enable African nations distribute essential medical supplies to rural communities [6].

This chapter explores the adoption of drones in African nations. It begins with describing drones. It discusses the uses of drones in Africa. It covers African nations that have adapted drones. It highlights the benefits and challenges of drones in Africa. The last section concludes with comments.

4.2 What Is a Drone?

Drones are autonomous robots that fly in the sky. They may also be regarded as pilotless aircrafts that were initially used by the military, but are now used for scientific and commercial purposes. The word "drone" was coined due to the similarity of its sound to a male bee. Drones are pilotless aircraft and are formally known as either unmanned aerial vehicles (UAVs) or unmanned aircraft systems (UASs). Drones are also called "remotely piloted vehicle" or "unmanned aerial systems." Drones were first used in the 1990s by military organizations. The notion of drones began around 1918 when the US Navy commissioned Charles Kettering built a militarized UAV. Their original use was to take strategic pictures for the military. From the beginning of the twenty-first century, civil activities of drones started to get more attention.

Drones are classified into different ways: size, weight, flight time, commercial or military, and cost. The US Federal Aviation Administration (FAA) defines consumer and commercial drones as those that weigh <1.0 lb (0.45 kg) with approximately a maximum of 500 m altitude and 2 km range from the base operator. A drone is a pilotless aircraft that operates through a combination of technologies, including computer vision, artificial intelligence, object avoidance tech, automation, robotics, and miniaturization.

Drones are becoming part of the tool kit for violent actors across Africa. As African nations continue to sort out the regulations and systems needed to support drone activities, drone technology promises to advance in Africa at an astounding pace. Drones were previously restricted to military operations, inspections, surveys, and deliveries. Today, drones are being used to solve some socioeconomic and environmental problems. The drone technology is bringing innovation into people's daily lives. Drones cost a fraction of what it takes to purchase and maintain manned aircraft and they provide a superior aerial advantage to any manned aircraft. Nearly every country on the continent uses drones in some form. In most cases, those uses are peaceful, from delivering healthcare supplies to remote communities to monitoring wildlife preserves under threat from poachers. Drones have become a force multiplier for security agencies across the continent.

4.3 Use of Drones in Africa

In spite of its poverty and poor infrastructure, sub-Saharan Africa has led the way in the adoption of mobile banking and healthcare, and now medical drones [7]. Necessity is the mother of invention; the rapid adoption of drone technology in sub-Saharan Africa is exemplary. The drone technology caught on so quickly in developing nations such as Rwanda, Tanzania, and Ghana with poor roads and lack of accessibility. The United Nations has used drones to drop condoms over rural parts of Ghana, where a fraction of women have access to contraceptives to prevent the transmission of sexually transmitted diseases. Threats of criminality, security breaches, and loss of privacy make drones too risky in developed countries. The World Economic Forum in partnership with Zipline and the World Bank has been raising awareness for using drones in Africa and beyond [8]. Tanzania is planning to have the largest drone healthcare delivery operation in the world. In spite of the fast adoption of drones in the continent, drones are manufactured elsewhere, e.g., United States and China. Drones are used in the following ways in Africa [8]:

– *Healthcare:* Drones have the transformative potential in revolutionizing healthcare delivery in Africa by addressing some of the continent's most pressing challenges. Drone delivery in healthcare has the ability to decrease costs, protect supply chain integrity, get around geographic restrictions, and speed up and improve delivery. Zipline is collaborating with governments of various countries to create a convenient delivery service to improve healthcare across the globe. The various applications of drones in healthcare delivery include delivering vaccines, medications, blood samples, diagnostic tools, and medical personnel to remote locations in a timely and cost-effective manner. Drones can also be used to spray larvicide in the swamps to fight against malaria. Delivering medical supplies to remote and underserved African areas poses significant challenges due to poor road infrastructure, long distances, difficult terrains, limited healthcare facilities, extreme weather conditions, and geopolitical factors that impact healthcare access. Drone technology presents a promising solution to overcome these challenges. Drone-enabled telemedicine services have emerged as a promising solution. Drones offer cost-effective solutions for transporting medical supplies and samples in Africa. Drone technology is capable of bridging the gap between remote areas and healthcare services, ensuring equitable access to medical care [9, 10]. Figure 4.2 shows a typical delivery of medical supplies [11].

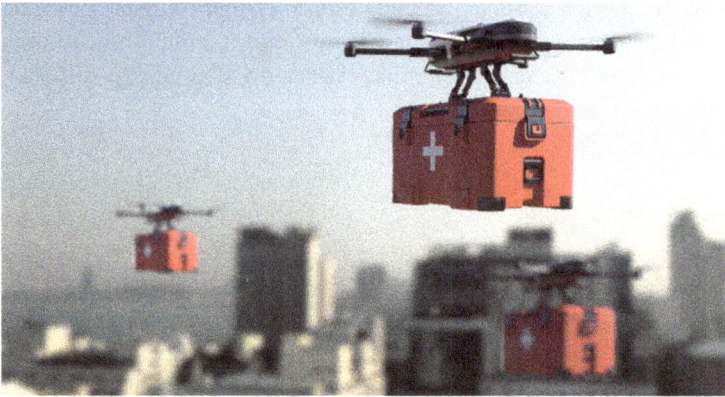

Figure 4.2: A typical delivery of medical supplies [11].

– *Agriculture:* Drones are essential for the modernization of African agriculture, as they bring more efficiency, precision, and reliability at a much lower cost. Farmers can use drones to monitor their fields more efficiently and determine much faster where and when they need to spray insecticide, water the fields, or any other necessary action. Drones are being adopted in Africa in crop agriculture due to the gradual transition from traditional farming to modern methods. All of this helps to save farmers' money and time and enhance their crop quality, yields, and profits on those yields.

– *Warfare:* Most media focus has been on the use of military drones by conventional armies and their impact on international humanitarian law. Drones now constitute the new weapon in African warfare. Military drones are rapidly changing the face of warfare. As an indispensable tool in modern warfare, drones are proliferating in

Africa, upsetting the balance of power on the battlefield. As shown in Figure 4.3, Ethiopian police officers take part in a drone piloting training program organized by China [12]. Across Africa, a flood of cheap imported drones is fueling conflicts. African armies and rebel militias see the drones as a highly useful and attractive weapon of warfare. They are turning to their affordable drones to fight insurgencies. The drones are often eagerly supplied by authoritarian states, including China, Iran, or Turkey [13]. One of the countries at the forefront of this new arms race is Nigeria, which has been using drones for surveillance and reconnaissance operations against Boko Haram. Other African countries that have acquired military drones include South Africa, Kenya, Morocco, Tunisia, and Algeria. They are all using drones for military purposes.

Figure 4.3: Ethiopian police officers take part in a drone piloting training program [12].

4.4 Adapting Drones in African Nations

Some countries, like Rwanda, have already developed innovative guides for governments to effectively regulate the use of drones, while other nations have created research and testing programs. Nearly every country on the continent uses drones in some form for military and nonmilitary purposes. We consider the following nations as typical African nations adopting the drone technology:

– *Rwanda:* This country was the first in the world to use drones to deliver blood and essential medicines to rural hospitals. Rwanda is among a group of early adopters of transport drones for healthcare systems in sub-Saharan Africa, with the first drone airport established in Rwanda. In 2016, the tech company Zipline, a company based in San Francisco, California, signed a deal with the Rwandan government for the delivery of logistical services in order to support the provision of healthcare in remote, infrastructurally disconnected areas of the country. The partnership between the Silicon Valley company Zipline and the Rwandan government gave birth to an innovative device, whose impact on citizens' health is crucial. Figure 4.4 shows the World Bank president, Jim Yong Kim, hailing Rwanda's use of drones in healthcare delivery [14]. The drone is operated by Zipline and is focused on lifesaving medical supplies. Zipline argues that minimizing waste in the medical system will help the drones pay for themselves. Zipline plans to expand its operations to more African nations. About 83% of Rwandans live in rural areas and the country is mountainous. Traditionally, when remote hospitals needed blood, it came by road. The drones have shuttled blood to rural, mountainous areas for years. The drones have also reduced the quantity of blood that expired and went to waste [15].

Figure 4.4: World Bank president hails Rwanda's use of drones in healthcare delivery [14].

– *Nigeria:* Many African countries such as Nigeria are adding drone technology to their arsenals. Nigeria deploys drones at border crossings to increase security along its porous frontier and to identify potential threats. The terrorists that Nigerian soldiers are looking for likely are watching them with drones of their own. Terrorist groups such as Boko Haram and its offshoot, the Islamic State West Africa Province, use drones for intelligence-gathering. The psychological impact of a weaponized drone in the hands of an insurgent is perhaps as important as its physical impact. Many of drones are made and marketed by China, and they have the potential to be weaponized by extremist groups. Nigeria also makes its own drones. Figure 4.5 shows

Nigerian military leaders and officials deploying drones for border security [16]. Experts say Nigeria needs to lead the regional effort but that the nation itself lacks a coherent policy to combat armed groups' use of the drones.

Figure 4.5: Nigerian military leaders and officials deploy drones for border security [16].

– *South Africa:* South Africa has become another major actor in the drone industry. South Africa manufactures drones for sale to other countries and has considered deploying them along its 5,244-km border. The flying of drones in the South African airspace had been unregulated and essentially illegal. South Africa is a member of the International Civil Aviation Organization (ICAO). It has seen the recent founding of the Drone Council to facilitate the expansion of the drone industry, with one of its objectives to promote a more balanced involvement of women. South Africa was the first African country to legalize commercial drones [17]. There are also a multitude of examples of champions for "drones for good," bringing the benefits of drone technology to farmers, town planners, conservationists, and disaster management – many of whom are women [18]. Fishing by drone has really taken off, and the United States has become the biggest market for the equipment developed in South Africa. Drones are being used to enable people to go deep-sea fishing, while standing on the beach. The South African National Defence Force is investing in drone technology. There is no policy on how drones are to be used and integrated into the military. Drones in South Africa could impact the essence of the African value of Ubuntu. The South African Civil Aviation Authority (SACAA) has now collaborated with the drone industry and formulated regulations to deal with this rapidly expanding industry. The regulations on flying recreational drones in South Africa are [19]:
- Fly at or below 400 feet.
- Keep your drone within sight.
- Never fly near other aircrafts or near any airports.

- Never fly over groups of people.
- Never fly over stadiums or sports events.
- Never fly near emergency response efforts such as fires.
- Never fly under the influence.
- Be aware of airspace requirements.
- Fly/operate Remotely Piloted Aircraft (RPA) or toy aircraft in a safe manner, at all times.
- RPA or toy aircraft should remain within the visual line of sight at all times.
- Fly/operate RPA in daylight and clear weather conditions.
- Inspect the aircraft before each flight.

– *Ghana:* The Ghana government and Zipline deployed the COVID-19 vaccine to hard-to-reach areas within Ghana. To facilitate the delivery, Ghana's Zipline branch has four distribution centers hosting a fleet of 30 fixed-wing drones that are serving as a drone airport and a medical supply warehouse. The drones can deliver the vaccines as far as destinations that are 22,500 km away from the distribution center. So far, the Zipline has distributed over 1 million vaccine doses in Ghana in over 50,000 deliveries.

– *Kenya:* Swarms of millions of locusts have decimated crops in many parts of Africa. In Western Kenya, drones have been used to spray and kill the swarms, as typically shown in Figure 4.6 [20]. Each drone has the capacity to cover 22 acres per hour, operate for up to 10 h a day, and carry up to 16 L of spray solution for every spray run. The drones can target otherwise inaccessible areas like roosting locusts atop trees. Drones not only have the potential for surveillance but for control operations targeting small swarms. Kenya is also engaged in spirited attempts to digitize its land records. In addition to digitization of existing records, Kenya is increasingly using drone technology in support of land mapping, particularly in remote areas, in order to enhance the accuracy of the data in the registries. The data obtained through the use of drones could facilitate more accurate delineation of boundaries among land holdings [21].

– *Niger:* In 2013, the United States expanded its drone operations in the middle of the Sahel from a base in Niamey in Niger. A number of drones and 100 military personnel were stationed in Niger "to promote regional stability in support of US diplomacy and national security and to strengthen relationships with regional leaders committed to security and prosperity." The drones fly in Niger, Mali, and Libya, and share broad patterns of human activity with French forces and other US partners who are fighting "terrorism." The US and the Nigerien government were worried that the foreign deployment of drones would trigger a backlash among civil society, as foreign military interventions are a sensitive issue. Therefore, both the United States and Niger have remained vague about why, when, where, and how drones would be deployed exactly. The President of Niger, Issoufou Mahamadou, welcomed the drones because he was "worried that the country might not be able to defend its borders from Islamist fighters based in Mali, Libya or Nigeria" [22].

Figure 4.6: Drones used to spray and kill swarms [20].

4.5 Benefits

In many African countries with poor road infrastructure, drones are becoming an efficient way of providing logistical services for the delivery of supplies in rural areas. Several African companies are currently using geographic information system (GIS) drones to collect aerial photos that are clearer than satellite photos. Drones are reshaping the way militaries from Morocco to South Africa protect their citizens, monitor their coastlines, combat wildlife, and more. Drones are being more and more developed on the continent because they offer advantages adapted to the local reality: the vastness of the territories, low costs of use, rapidity of airway transportations, etc. Drone technology is already being applied to solve the challenges of delivering medical supplies to rural and hard-to-reach areas that are cut off from the main transportation systems in major cities. Other benefits of drones in Africa include the following [23]:

1. *Low Cost:* Drones range in cost from $99 to tens of thousands of dollars. The low cost of drone technology relative to a full-scale naval ship or air force jet means that cash-strapped militaries can add capabilities on a shoestring. Drones also operate for limited amounts of time before having to refuel. That means that they can cover vast distances effectively.

2. *Humanitarian Aid and Medical Supply:* Perhaps the most exciting development for humanitarian aid workers in Africa is the use of drone technology in emergency medical situations. Drones are proving to be essential delivery vehicles for humanitarian aid and health supplies. In less than 10 years, drones or UAVs have transformed healthcare within sub-Saharan Africa, where they deliver life-saving medicine, vac-

cines, and blood to solve last mile hurdles. The drone is a powerful tool for governments to address the equity gap, improving access to primary healthcare to the underreached. As shown in Figure 4.7, a woman packs a box of vaccines to be delivered by a Zipline drone, in Ghana [24].

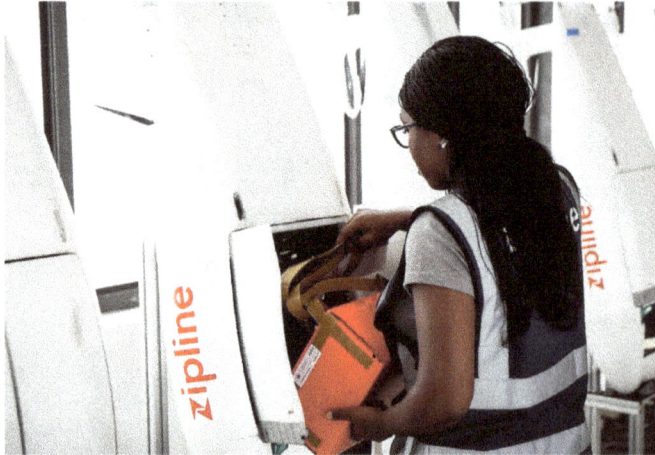

Figure 4.7: 3. A woman packs a box of vaccines to be delivered by a Zipline drone [24].

3. *Hobbyist Drone:* The ease with which shop-bought or hobbyist drones can be acquired across Africa suggests that indigenous innovation may appear to be more likely than direct technology transfer.

4. *Prosperity:* Today, drones are being used to solve some socioeconomic or environmental problems. The drone industry in Africa is taking off and it is evolving into a massive enterprise. People have begun to recognize the many benefits of using drones for commercial and noncommercial purposes. Drone technology can potentially solve some of the biggest challenges facing the region and be a major catalyst toward a future of prosperity, peace, and independence. South Africa has become a major actor in the drone industry. South Africa manufactures drones for sale to other countries.

5. *Military Use:* Drones were initially developed as military tools for surveillance and targeted air strikes. They have become indispensable in modern warfare. Drones are reshaping the way militaries from Morocco to South Africa protect their citizens, monitor their coastlines, combat wildlife poaching, and more. Nations are less transparent about their drone use when it comes to their militaries. Critics say that lack of transparency leads to concerns about whether drones are being used in accordance with international human rights standards and traditional rules of engagement. An uncountable number of drones are increasingly monitoring and impacting lives all over North Africa. An arms race may be in the offing. Drones have also proved effective in

traditional conflicts, particularly the civil wars in Ethiopia and Libya, where opposing sides are more easily identifiable. But deploying them against jihadists and insurgents in northern Nigeria, Somalia, and across the Sahel has been highly problematic.

6. *Agriculture:* Drones are important for the modernization of African agriculture, as they bring more efficiency, precision, and reliability at a much lower cost. They help to optimize agricultural yields because they significantly reduce production costs. Drones can help enhance and optimize crop yields via aerial monitoring of fields. Drones can now hover over fields of vegetables and grain, such as maize, sweet potato, and rice, with special infrared sensors that can collect the aerial data which farmers and governments can use to better understand and predict crop yield, assess crop health, and keep the weed cover at bay. Farmers can use drones to monitor their fields more efficiently and determine much faster where and when they need to spray insecticide, water the fields, or any other action that may be necessary. Drones can be used for precision agriculture to help farmers decide when and where to apply fertilizer or irrigate crops.

7. *News Gathering:* Journalism has rapidly evolved over the years due to advancements in technology that have produced new tools and techniques for news gathering. "Drone journalism," the use of drones for news gathering, is on the rise across the African continent. Drones can be useful tools for obtaining aerial images and news footage of areas that would otherwise be difficult to cover. Although drones are a relatively cost-effective method of gathering news, they can still be prohibitively expensive for smaller newsrooms, particularly the high cost of training journalists on how to use them safely.

4.6 Challenges

In spite of their capabilities, drones have limits. Establishing drone networks requires time, investment, and coordination. Some leaders refuse to spend the extra money to train operators, opting to have them figure it out on their own. That "do-it-yourself" approach can result in drones lost on the bottom of the ocean or lying in a pile of wreckage on the ground. Other times, the equipment goes nowhere at all. Other challenges of drones in Africa include the following [25]:

1. *Privacy and Safety:* An area of major concern is privacy. Despite the usefulness of drones, there are legitimate concerns over safety and privacy associated with their use. Flying drones with cameras, scanners, and sensors could allow unscrupulous individuals to anonymously collect and record sensitive or damaging information on civilians, businesses, and other organizations. Aerial data must be regulated because it is related to the protection of privacy. Security must be reliable in all possible ways. Due to the proliferation of easy-to-obtain drones, thousands of African civilians have

been killed by drones in recent times. Markets, schools, hospitals, weddings, funerals, and civilian vehicles have all been hit by poorly targeted drone strikes.

2. *Terrorist Organizations:* The proliferation of drones has allowed terrorist organizations to execute operations that were once exclusive to military and state organizations. Drones have become a vital tool valued by security forces and terrorist groups alike in recent years. Security experts in Africa are raising concerns about the growing use of drones by terrorist groups and the readiness of government forces to match their sophistication. The use of drones by violent extremist organizations highlights their consistent technological advancements to enhance their operational capabilities. The terrorists that Nigerian authorities are looking for likely are watching them with drones of their own. Terrorist groups such as Boko Haram and its offshoot, the Islamic State West Africa Province, use drones for intelligence-gathering. Mozambique's Interior Minister Amade Miquidade reported that UASs have been deployed by militant Islamist groups in Cabo Delgado Province, where a Southern African Development Community stabilization force has recently been authorized. Mozambique's armed forces have been battling these militant groups. The Mozambican experience mirrors other reports emerging from Africa.

3. *Threat:* Drones pose a serious and evolving threat to security and stability on the continent. The use of drones for suicide bombings or kamikaze attacks against targets poses the biggest threat. By attaching explosives or other payloads to drones, terrorists can turn them into flying bombs that can be remotely detonated or programmed to crash into targets. By using drones equipped with cameras and sensors, terrorists gather information about their targets, monitor enemy movements, and scout for vulnerabilities. This helps them plan and coordinate their attacks more efficiently.

4. *No Regulations:* Regulation of civilian airspace in Africa is the responsibility of the National Civil Aviation Authorities (NCAAs), which control both the development and enactment of regulations. Although the aviation authority has not yet formalized its regulations, it has developed ad hoc requirements for drone operators. Many African countries are still struggling to put the necessary regulations in place to support drone operation. However, some countries, like Rwanda, have already developed innovative guides for governments to effectively regulate the use of drones.

5. *Airspace Management:* To realize the full potential of drone technology safely and effectively, airspace management is critical. In Israel, the Israel National Drone Initiative (INDI) in 2019 promoted the use of drones for the betterment of society. The INDI aims to seamlessly integrate drone and airspace management technologies necessary for drones to operate safely. Airspace management is critical to safely and effectively realizing drone technology's full potential. The INDI promotes the use of drones for societal benefit, showcasing all necessary technologies.

6. *Technical Skills:* Another challenge is the technical skills required to operate the drones. While some commercial drones are relatively easy to use, others may require more sophisticated training to fly and control them. Building a drone network requires a local workforce with technical expertise, including pilot licenses.

4.7 Conclusion

The age of drones has arrived and drones are here to stay. Autonomous flying is being applied in several industries ranging from transportation to law enforcement and defense. The drone industry in Africa is taking off and it is evolving into a massive enterprise. The drone technology can enable African countries to flourish and prosper. The addition of thousands of new technology and drone-related jobs to the global economy presents a unique growth opportunity for African youths.

Drones can be a force for good and does not understand geographical barriers. The recent proliferation of drones throughout the African continent offers a ray of hope to the over 1 billion people living on the continent. Perhaps the most exciting development for humanitarian aid workers in Africa is the use of drone technology in emergency medical situations. The Drone & Data Academy will prepare young Africans with the skills that will be needed to join the new technology workforce [26]. The capacity for drone warfare in a continent experiencing various armed conflicts is certain to become relevant. For more information about the use of drones in Africa, one should consult the books in [27–30] and the following related journals devoted to drones:

– *Journal of Unmanned Vehicle Systems*
– *Journal of Intelligent Robot System*

References

[1]   "Flying drones in Africa: The challenges and the opportunities," https://www.shearwater.ai/post/flying-drones-in-africa-the-challenges

[2]   K. Haula and E. Agbozo, "A systematic review on unmanned aerial vehicles in Sub Saharan Africa: A socio-technical perspective," *Technology in Society*, vol. 63, November 2020.

[3]   P. Kardasz et al., "Drones and possibilities of their using," *Journal of Civil & Environmental Engineering*, vol. 6, no. 3, 2016.

[4]   "Drones and blood transportation: Will drone impact society?" https://mbamci.com/drones-and-blood-transportation/

[5]   P. Tilsley, "Drones increasingly used in Africa to save people's lives, deliver blood samples to labs," August 2019, https://www.foxnews.com/world/drones-africa-save-lives-blood-samples-labs

[6]   M. N. O. Sadiku, U. C. Chukwu, and J. O. Sadiku, "Drones in Africa," *Innovative: International Multidisciplinary Journal of Applied Technology*, vol. 2, no. 4, 2024, pp. 77–87.

[7]   B. McCall, "Sub-Saharan Africa leads the way in medical drones," *World Report*, vol 393, January 2019, pp. 17–18.

[8] R. Sengupta, "Drones deliver medicines in Africa," June 2019, https://www.downtoearth.org.in/news/health-in-africa/drones-deliver-medicines-in-africa-64832

[9] "The drone industry in Africa," https://www.do4africa.org/en/the-drone-industry-in-africa/

[10] G. Olatunji et al., "Exploring the transformative role of drone technology in advancing healthcare delivery in Africa; a perspective," *Annals of Medicine & Surgery* (London), vol. 85, no. 10, October 2023, pp. 5279–5284.

[11] J. Orisakwe, "Rise of drone delivery service for medical supplies in Africa," August 2023, https://insights.omnia-health.com/management/rise-drone-delivery-service-medical-supplies-africa

[12] N. Hochet-Bodin, "Drones: A weapon system coveted by all African armies," December 2023, https://www.lemonde.fr/en/international/article/2023/12/30/drones-a-weapon-system-coveted-by-all-african-armies_6388927_4.html

[13] "Drones in Africa: Cheap UAV proliferation changes warfare and fuels surge in deaths," May 2024, https://northafricapost.com/77187-drones-in-africa-cheap-uav-proliferation-changes-warfare-and-fuels-surge-in-deaths.html#:~:text=Across%20Africa%2C%20a%20flood%20of,indiscriminate%2C%20thus%20increasing%20civilian%20casualties.

[14] "World Bank president hails Rwanda's use of drones in healthcare delivery," http://www.xinhuanet.com//english/2017-03/22/c_136148796.htm

[15] "Drones have transformed blood delivery in Rwanda," April 2022, https://www.wired.com/story/drones-have-transformed-blood-delivery-in-rwanda/

[16] "The dizzying possibilities of drones," https://adf-magazine.com/2023/09/the-dizzying-possibilities-of-drones/

[17] "South Africa: First African country to legalise commercial drones," https://trends.directindustry.com/project-19318.html

[18] "Trailblazing women in the African drone industry," https://womenanddrones.com/wd-africa/

[19] "Drone regulations – Southern Africa," https://akdmc.com/media/6684/ak-southern-africa-drone-laws.pdf

[20] "Pilot project using drones to control desert locust launches in Kenya," October 2020, https://blog.invasive-species.org/2020/10/26/pilot-project-using-drones-to-control-desert-locust-launches-in-kenya/

[21] P. Kameri-Mbote and M. Muriungi, "Potential contribution of drones to reliability of Kenya's land information system," *The African Journal of Information and Communication*, vol. 20, 2017, pp. 159–169.

[22] "Remote Horizons: Expanding use and proliferation of military drones in Africa," https://paxforpeace.nl/wp-content/uploads/sites/2/import/2021-05/PAX_remote_horizons_FIN_low res.pdf

[23] B. Odeba, R. Barnabas, and M. B. Daburi, "The use of technology in newsgathering," *International Journal of Research and Innovation in Social Science*, vol. 1, no. 1, January 2022.

[24] M. Leedom, "Drones deliver humanitarian aid in Africa," June 2024, https://www.thinkglobalhealth.org/article/drones-deliver-humanitarian-aid-africa

[25] "Flying drones in Africa: The challenges and the opportunities," https://www.shearwater.ai/post/flying-drones-in-africa-the-challenges

[26] "The African Drone and Data Academy," https://adda.cired.vt.edu/

[27] C. S. Rinehart, *Drones and Targeted Killing in the Middle East and Africa: An Appraisal of American Counterterrorism Policies.* Lexington Books, 2018.

[28] D. Soesilo and G. Rambaldi, *Drones in Agriculture in Africa and Other ACP Countries: A Survey On Perceptions and Applications.* CTA, 2018.

[29] A. Stokenberga and M. C. Ochoa, *Unlocking the Lower Skies: The Costs and Benefits of Deploying Drones Across Use Cases in East Africa.* World Bank Publications, 2021.

[30] F. E. Mbuya, G. Rambaldi, and H. R. Chaham, *Getting Drones Off the Ground in Africa.* CTA, 2017.

Chapter 5
Big Data in Africa

Data analytics is the future, and the future is NOW! Every mouse click, keyboard button press, swipe or tap is used to shape business decisions. Everything is about data these days. Data is information, and information is power. – Radi

5.1 Introduction

Africa is a continent that has 54 countries, with an area of 30,370,000 km^2 and 1.4 billion individuals as of 2021, subdivided into 5 major regions, like Northern Africa (with countries like Libya, Egypt, North Sudan, Algeria, Morocco, and Tunisia, as demonstrated) inhabiting the northerly region of Africa [1]. Due to its size and proximity to the equator, the climate of Africa ranges from the arid deserts in the north to dense rain forests in the south. The African continent is home to immense potential and abundant resources. It is home to some of the youngest populations in the world. It is marked by rich diversity, not only in terms of culture, language, and geography but also in the digital landscapes across the continent. The penetration of smart mobile phones has been a key driver of digital connectivity across Africa. It is well known that the African continent has been experiencing one of the most dynamic digital growths in the world for several years.

We live in an era of data. The world's most valuable resource is no longer oil, but data. The word "data" means information in a raw or unorganized form (such as alphabets, numbers, or symbols) that represent conditions, ideas, or objects. Data yields information, which yields knowledge, which yields wisdom.

Data is increasingly integrated into every aspect of our lives. It has become a high-value commodity. The Internet is being flooded with a huge amount of data, which is consolidating the big data (BD) concept. BD is an umbrella term for large volumes of structured and unstructured data that we encounter daily. It may be regarded as a collection of data from traditional and digital sources inside and outside an organization. BD is a fast-moving field that has taken the business world by storm. It is a pressing issue, particularly at a time when many are concerned about the role of information in political change. Organizations, both big and small, are opening up to the importance of data and the impact it can bring to organizations. The exponential growth of global data is not slowing down anytime soon [2]. Figure 5.1 highlights several sources of the BD deluge [3].

Application of BD has brought tremendous advancements in law enforcement and military intelligence, space science, aviation, banking, and in almost every aspect of human endeavor. BD is already revolutionizing operations in traditional industries

https://doi.org/10.1515/9783112211984-005

Whats's driving Data Deluge?

Mobile Sensors	Social Media	Video Surveillance	Video Rendering
Smart Grids	Geophysical Exploration	Medical Imaging	Gene Sequencing

Figure 5.1: Sources of the big data deluge [3].

by increasing efficiency. Recent research shows how resilience is enhanced through the use of BD analyses.

BD refers to vast volumes of data that are too large for ordinary computing devices to process. It is having a positive impact in almost every sphere of life, such as military intelligence, space science, aviation, banking, and healthcare. It has gained substantial interest among academics and business practitioners in the digital era. BD enables data-driven decision-making, which can lead to more efficient and effective economic policies, as well as new opportunities for investment and development. The opportunity to use open data and BD has become more popular in recent years. The African continent has been experiencing one of the most dynamic digital growths in the world for several years. BD has the potential to revolutionize the African economy. But Africa is usually plagued with the same problem over and over again, time lag. By utilizing new and innovative technological tools, African countries can unlock their economic potential and achieve sustainable growth [4].

This chapter examines the impact and prospects of BD in developing African nations. It begins with explaining the concepts of BD and BD analytics. It addresses the use of BD in Africa. It covers some African nations adapting BD. It highlights the benefits and challenges of BD in Africa. The last section concludes with comments.

5.2 Concept of Big Data

The act of collecting, storing, and analyzing data have been around for a very long time since Mesopotamia. What has changed, however, is the size and complexity of the data itself. Now, data is so large, fast, or complex that it is difficult or impossible to process using traditional methods. It is now being generated at a speed and volume never before experienced.

BD refers to a collection of data that cannot be captured, managed, and processed by conventional software tools. It is a relatively new technology that can help many industries, including the government. The three main sources of BD are machines, people, and companies. BD is essentially classified into three types: structured data, unstructured data, and semi-structured data. The different types of BD are depicted in Figure 5.2 [5].

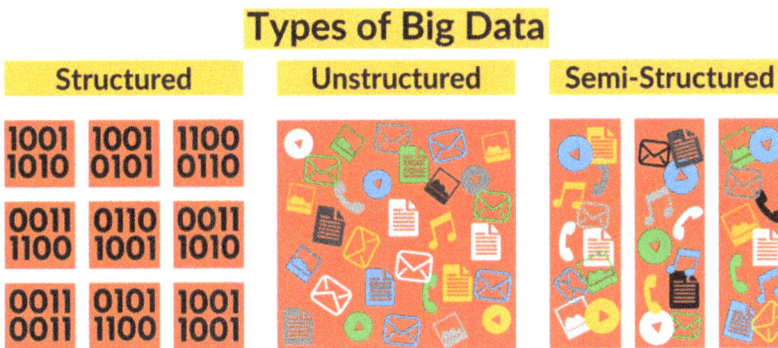

Figure 5.2: Types of big data [5].

BD can be described with 42 Vs [6]. The first five Vs are volume, velocity, variety, veracity, and value [7].

- *Volume:* This refers to the size of the data being generated both inside and outside organizations and is increasing annually. Some regard BD as data exceeding 1 petabyte in volume.
- *Velocity:* This depicts the unprecedented speed at which data are generated by Internet users, mobile users, social media, etc. Data are generated and processed. The volume of data is only increasing by the year in a fast way to extract useful, relevant information. BD could be analyzed in real time, and it has movement and velocity.
- *Variety:* This refers to the data types, since BD may originate from heterogeneous sources and is in different formats (e.g., videos, images, audio, text, logs). BD comprises structured, semi-structured, or unstructured data.
- *Veracity:* By this, we mean the truthfulness of data, i.e., whether the data comes from a reputable, trustworthy, authentic, and accountable source. It suggests the inconsistency in the quality of different sources of BD. The data may not be 100% correct.

- *Value:* This is the most important aspect of BD. It is the desired outcome of BD processing. It refers to the process of discovering hidden values from large data-sets. It denotes the value derived from the analysis of existing data. If one cannot extract some business value from the data, there is no use in managing and storing it.

On this basis, small data can be regarded as having low volume, low velocity, low variety, low veracity, and low value. An additional five Vs have been added [8]:
- *Validity:* This refers to the accuracy and correctness of data. It also indicates how up-to-date it is.
- *Viability:* This identifies the relevancy of data for each use case. Relevancy of data is required to maintain the desired and accurate outcome through analytical and predictive measures.
- *Volatility:* Since data are generated and change at a rapid rate, volatility determines how quickly data changes.
- *Vulnerability:* The vulnerability of data is essential because privacy and security are of utmost importance for personal data.
- *Visualization:* Data needs to be presented unambiguously and attractively to the user. Proper visualization of large and complex clinical reports helps in finding valuable insights.

In addition to the 10 Vs mentioned above, some suggest the following 5 Vs: venue, variability, vocabulary, vagueness, and validity. Figure 5.3 shows 10 Vs of BD [9]. The future of BD will bring more Vs. BD promises to drive economic growth and development in Africa. Figure 5.4 displays the big data readiness index (BDRI) for all African countries and shows the top ten performers [10]. Looking at the location of the top performers, coastal countries perform well, and neighbors can influence each other in terms of technology adoption. Looking at the rankings shows that coastal nations and islands perform best, with Rwanda being the only non-coastal country in the top 10.

5.3 Big Data Analytics

Big datasets can be staggering in size so that its analysis is daunting. Every day, data is growing bigger and bigger, and BD analysis (BDA) has become a requirement for gaining invaluable insights into data such that companies could gain significant profits in the global market. BD analytics can leverage the gap within structured and unstructured data sources. Once the BD is ready for analysis, advanced software programs such as Hadoop, MapReduce, MongoDB, Spark, Cassandra, Apache Storm, and NoSQL databases are utilized [11]. BD analytics refers to how we can extract, validate, translate, and utilize BD as a new currency of information transactions. It is an

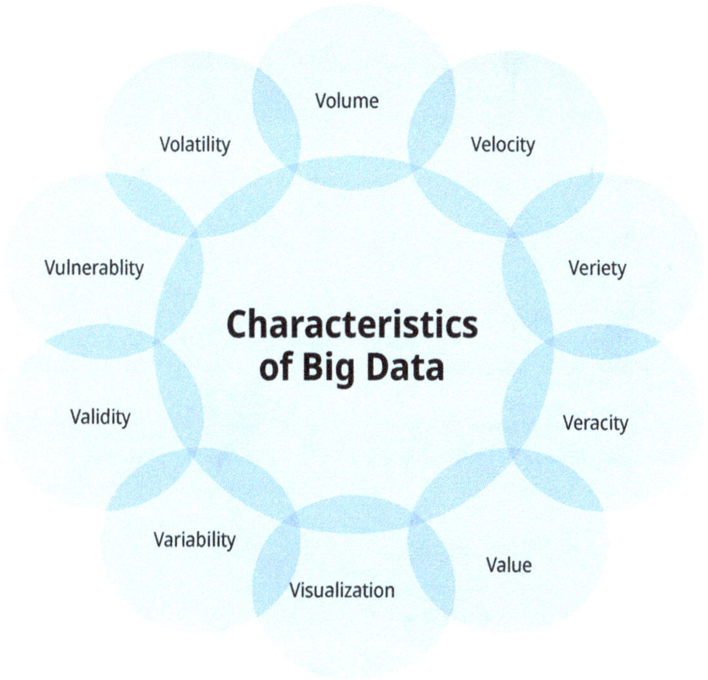

Figure 5.3: Characteristics of big data [9].

emerging field aimed at creating empirical predictions. Data-driven organizations use analytics to guide decisions at all levels [12]. BD and analytics continue to spark interest among scholars and practitioners worldwide.

In essence, big data analytics [13]:

- It refers to a huge amount of data that remains growing with respect to time.
- Its volume is so much that it cannot be processed or analyzed using normal data processing techniques.
- It contains data storage, data sharing, data mining, data visualization, and data analysis.
- The term is an all-comprehensive one, encompassing data frameworks, data, along with tools and techniques used to process and analyze the data.

The key to BD is analytics. BD analytics is capable of processing massive amounts of dirty data and extracting the gold information from it. It basically learns from data to predict the way individuals will behave in the future. It is on the verge of a major transformation. The process involved in analyzing BD is shown in Figure 5.5 [14]. Industries making investments in BD analytics include education, banking, retail, manufacturing, finance, healthcare, transportation, government, and the military.

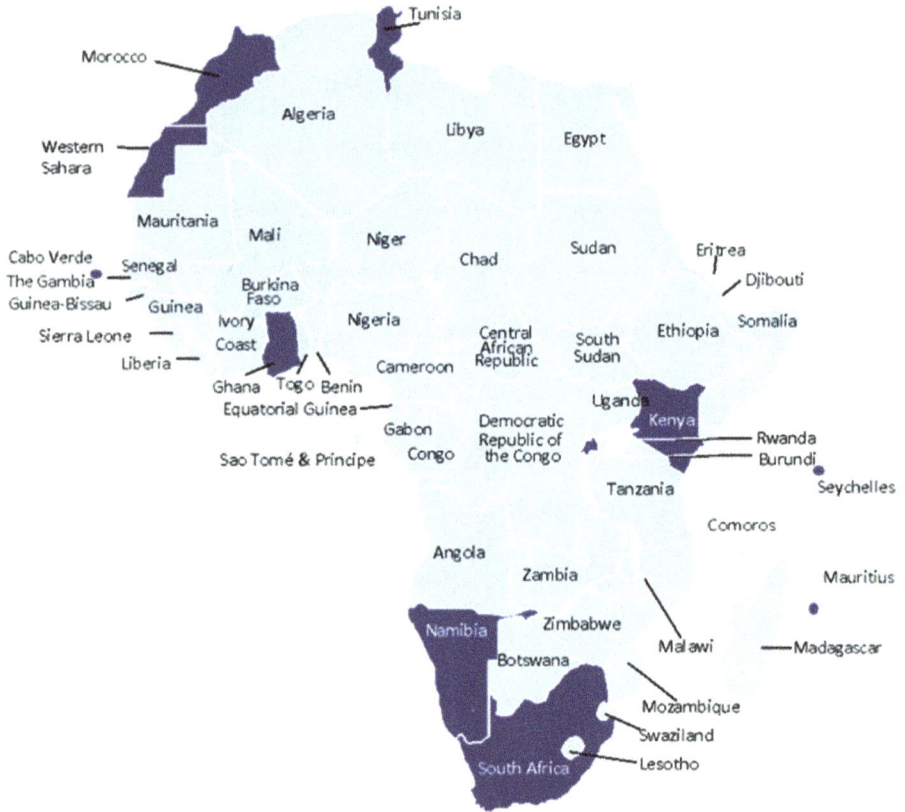

Figure 5.4: Big data readiness index (BDRI) for all African countries [10].

BD analytics has gained substantial interest among academics and business practitioners and promises to deliver improved operational efficiency, drive new revenue streams, ensure competitive advantage, and facilitate innovation. Analysts believe that by harnessing the power of BD, Africa can leapfrog into the future and realize its true potential as a leading player in the global economy. BDA also has enormous potential as a motor for the attainment of the 17 UN Sustainable Development Goals (SDGs).

5.4 Use of Big Data in Africa

The deployment of cutting-edge digital technologies, such as artificial intelligence (AI) and BD, has the potential to revolutionize various sectors, including healthcare, education, agriculture, and finance, and drive sustainable development and economic growth across the continent [15]. If Africa gets it right and accelerates the move to

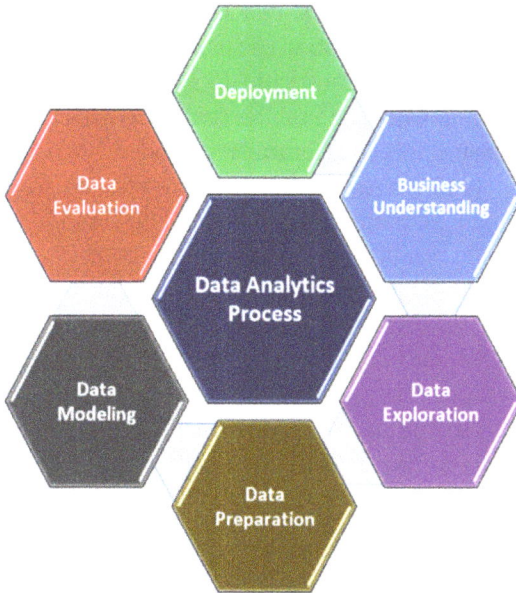

Figure 5.5: The process involved in analyzing big data [14].

adopt these technologies, the benefits are immense. Skills and job acceleration, business growth, trade, and innovation are potential benefits. BD is an all-inclusive approach to managing, processing, and analyzing huge volumes of data for providing insights, so that we can have a better advantage when making decisions. It promises to change the way businesses operate in healthcare, manufacturing, agriculture, finance, space, and other industries. Governments across Africa are recognizing the transformative potential of BD and implementing initiatives to harness its benefits. We consider how BD is being utilized in some selected African nations.

– *Healthcare:* BD is a growing force in healthcare. Healthcare BD comes from different sources such as sequencing data, electronic health records, biological specimens, biomedical data, patient-reported data, biomarker data, medical imaging, large clinical trials, surveillance data, diseases' registries, etc., while other datasets may be derived from wearable devices and other digital footprints. Healthcare systems in the developed world are recording some breakthroughs due to the application of BD, and it is important to research the impact of BD in developing African nations. Although healthcare systems in Africa are relatively behind the rest of the world, the technologies to amass BD, such as the Internet and mobile phones, are already in use in Africa. In the past, records were stored as hard copies. Now, records are digitized, leading to large volumes of data being generated. There is no doubt that the era of digitized records has helped improve the quality of healthcare delivery, reduce healthcare costs, improve efficiency in disease surveillance, and enhance public health management. The availability of BD in the healthcare sector presents an opportunity to discover relationships, patterns, and trends within the data. Due to the potential benefits of BD in healthcare, analytics of BD in healthcare is now an emerging discipline in medical science. It is logical to assume that BD has a significant role to play in improving the health sector of developing regions of the world, such as Africa. The poor state of healthcare in Africa does not mean that healthcare in Africa will remain in a pathetic state forever; the continent has great prospects for BD applications in healthcare. Although the application of BD is still in its infancy in Africa, research provides insight that BD analytics is emerging in Africa and can be a big weapon to improve healthcare and to end many diseases plaguing the continent [16]. A relational philosophy counts as African if it is informed by values that are more prominent on the continent. The African philosophy of *Ubuntu* can usefully influence BD practices in ways that address this challenge without undermining its benefits, since *Ubuntu* emphasizes harmonious relationships [17]. Another development in the use of BD in healthcare in Africa is digital surveillance in tracking epidemics.

– *Agriculture:* Agriculture is a cornerstone of many African economies, and BD has the potential to revolutionize the sector. A better understanding of the continent's agricultural potential will allow African countries to make better informed decisions on where to invest in agriculture. Data analytics can optimize crop yields, improve resource management, and provide valuable insights for farmers. South Africa's agriculture sector is benefiting from the amalgamation of AI and BD. Precision agriculture techniques, guided by AI, enable farmers to maximize crop yields while conserving resources.

– *Finance:* BD plays a crucial role in advancing financial inclusion in Africa. Mobile banking and digital payment platforms utilize data analytics to assess creditworthiness. The inclusive approach empowers unbanked populations, fostering economic participation and growth. Banks and financial institutions are harnessing AI for fraud detection, risk assessment, and customer service. This has led to more secure transactions and improved customer experiences. BD has the potential to completely transform African economies and finance, but to do so, an educated workforce in BD analytics is needed. In order to control the cost of their investments, finance departments are putting considerable effort into retrieving and collating data from disparate sources. Analytics is used to calculate financial statements and determine hypothetical results under different scenarios.

– *Education:* Education is key to development and innovation. Educational systems in Africa are witnessing growth and the embracing of technological trends. The educational sector creates tremendous amounts of data using technology. Education today faces increased competition, assessment, collaboration, and regulation sectors across the globe. There is a need for tools to harvest information from a pool of heterogeneous and homogeneous data that leads to accurate decision-making by the top management of the educational sector. In the context of educational BD, a high level of intelligence is required in data analysis and the development of predictive models. In Africa, with many institutions of learning, academic analytics can be used to harness generated data for growth, research, and innovation. Although much has been achieved in the area of primary education in sub-Saharan Africa, there is little progress in the training of primary school teachers and in skill acquisition in the region. Political uncertainties exist in most African nations, obstructing the free running of the educational system in such nations [18].

– *Government:* The role of BD in policymaking is a significant concern in our information age. To realize value from BD, governments must strengthen technical and legal frameworks to access and use data responsibly. Governments have an opportunity to harness BD solutions to improve productivity, performance, and innovation in service delivery and policymaking processes. In the public sector, BD refers to the use of non-traditional data sources and data innovations to make government solutions more responsive and effective. The potential for BD to transform government is vast. People interact with government services every day in health, employment, education, business, elections, etc. Society is eager to use BD to make public service delivery equally smart, responsive, and personalized [19].

– *Space:* African countries are increasingly employing space technologies in addressing several challenges. Satellite data, or space data, is a subset of BD, and its services have become increasingly affordable. While obtaining satellite data is imperative, analyzing it properly and knowing how to handle this large volume of data is important. Space data has the potential to revolutionize how we understand a wide array of industries and environmental phenomena. It can be used in farming to monitor factors which influence crop yield. In real estate, areas prone to flooding or sinkholes can be accurately identified using space data. In retail, foot traffic around shopping centers can be monitored in real-time. Space data can also be used to track refugee movement [20].

5.5 Adapting Big Data in African Nations

The African continent has not yet taken full advantage of the prospects presented by BD. BD has the potential to alter African economies. The development of BD and analytics is an economic, societal, and sovereign necessity for the entire African continent. Let us explore how BD is revolutionizing African nations.

– *South Africa:* South Africa is a middle-income country with a population of 52 million, 53% live below the poverty line, 24% are unemployed, and 11% live with HIV/AIDS. South Africa and other collaborating countries in Africa will need to prioritize the management, analysis, publication, and curation of "Big Scientific Data." BD has emerged as a valuable asset for South Africa. The country is generating an immense amount of data daily, providing a treasure trove of insights waiting to be discovered. BD analytics is helping uncover patterns, correlations, and trends that were previously difficult to discern. This aids in making data-driven decisions across sectors, from urban planning to healthcare management. South Africa will have to cope with an increasing deluge of scientific data. The FAIR Data Principles are intended as a guide to making data Findable, Accessible, Interoperable, and Reusable (FAIR). With the deluge of research data expected from the new experimental facilities in South Africa, the problem of how to make such data FAIR takes center stage [21]. Discovery, a major South African business, uses website logs, Internet clickstreams, social media activity, and mobile-phone data for insights to effectively understand customer needs and cross-sell more products to customers. South Africa and China have held two workshops on the BD challenge in Astronomy under Chinese-South African collaborations. The fusion of AI and BD is at the forefront of transforming South Africa's financial, healthcare, agricultural, and other sectors.

– *Zambia:* PATH, a Seattle-based non-profit dedicated to global health, has worked in Zambia since 2005 in conjunction with the Ministry of Health. Their main goal is to assist the National Malaria Elimination Center, including data administration, research, community engagement, and routine and campaign malaria operations. The group created the Viz Alerts software, which is an open-source Python script made to find data gaps, notify nearby health workers of those gaps, and incorporate missing data when it is supplied by health workers [22].

– *Nigeria:* Initiatives like Data Science Nigeria (DSN) are responding to the brain drain challenge and stepping up to position Nigeria as a prominent player in the global data science outsourcing market. DSN's focus on enhancing the BD and machine learning ecosystem aims to secure a substantial portion of this growing sector. DSN is not only emphasizing the transformative potential of data but is also striving to pivot Nigeria's economy from oil reliance to a data-driven model. This aligns with the global trajectory where data science is increasingly recognized as critical, highlighting the importance of retaining such talent within the continent [23]. A Nigeria-based agritech startup is currently working on launching a satellite-enabled BD analytics dashboard called "Village Chief."

– *Ghana:* Due to structural inconsistencies, statistical authorities in Ghana took 17 years to adopt the UN national accounting system. Ghana-based Farmerline provides best-practice information to farmers on managing farms and increasing yields.

– *Uganda:* Some interesting examples of how BD analytics are being applied to assist development include using machined roof counting to measure poverty in Uganda. The National Population Secretariat was established to provide leadership for data collection and management in Uganda.

– *Rwanda:* Given the relevance of the BD hub to Rwanda and Africa, the National Institute of Statistics of Rwanda (NISR) realizes that partnerships and collaboration with other institutions and organizations are needed to achieve its mission. NISR is ensuring that all personal data is used in a lawful manner. In March 2020,

Rwanda signed a memorandum of understanding (MoU) with the United Nations to operationalize this hub, which was expected to start before the end of 2020. The MoU agreed that Rwanda would host the only BD regional hub for Africa. Many areas were identified where BD analytics will be used to provide useful insights for evidence-based decision-making. Rwanda is keeping pace with this agenda of BD. To build the national capacity for BDA, Rwanda has three universities offering courses and short-term trainings related to BD [24].

5.6 Benefits

By itself, data is quite inconsequential. Much like gold, data needs to be mined to extract its value. There is a growing adoption of BD in various fields such as engineering, life sciences, healthcare, business, behavioral studies, online and offline commerce, education, and politics. The data revolution in Africa is driven by several key industries that recognize the potential of data analytics for growth and development. These industries are leveraging data to make informed decisions, enhance efficiency, and create innovative solutions. With the right analysis tools, this data can even be used to predict future buying decisions. Other benefits of BD in Africa include the following [25, 26]:

- *Poverty Eradication:* Analysis of BD has the potential to provide solutions for those working on poverty eradication. While the use of BD is fairly new terrain, an increase in the use of mobile technologies and social media in Africa, combined with investment in BD technology, makes strategies to combat poverty on the continent easier to formulate and implement.
- *Business Transformation:* We are living in a digital age where data has become the new currency, and harnessing its potential is crucial for businesses worldwide. As data continues to surge in businesses across Africa, the transformation of this wealth of information into actionable insights becomes a matter of utmost importance. BD analytics empowers businesses to make data-driven decisions. It is poised to revolutionize African businesses. By analyzing customer data, businesses can personalize their products and services, improving customer satisfaction and loyalty. African information and communications technology (ICT) companies are playing a pivotal role in enabling this transformation by providing the necessary infrastructure, expertise, and solutions tailored to the local market. Embracing BD analytics is a necessity for staying competitive and relevant in this rapidly evolving business landscape.
- *Business Growth:* BD has the power to underpin new waves of growth and innovation and bring Africa into the data economy. If Africa gets BD right, the benefits will be far-reaching and will include job creation, business growth, trade, and innovation. BD can be used to improve public health, prevent disease outbreaks, and identify areas in need of economic development. BD can help African countries improve their governance, reduce corruption, and increase transparency in

public services. BD promises to improve government service delivery, complement official statistics, and facilitate development in a range of sectors. The private sector has an important role to play in fostering the use of BD in Africa.

– *Skill Development:* African ICT firms understand the unique challenges faced by local businesses. These companies are playing a pivotal role in training a workforce skilled in data analytics. They provide training programs and certifications to ensure businesses have access to data professionals.

– *Data Security:* African ICT companies prioritize data security, ensuring that sensitive information remains protected. They implement state-of-the-art security measures to safeguard against data breaches. The more data a company keeps, the more data it has at risk.

– *Financial Inclusion:* BD plays a major role in transforming the financial sector and in driving financial inclusion and innovation in Africa. African financial institutions are using data analytics to assess credit risk, detect fraud, and personalize customer experiences. Mobile banking and digital payment platforms generate vast amounts of transaction data that can be analyzed to improve services.

– *Healthcare:* The private sector has an important role to play in fostering the use of BD in Africa. A popular sector where BD analytics is being used is healthcare. Data analytics is transforming healthcare in Africa by improving patient outcomes, disease monitoring, and resource allocation.

– *Agriculture:* The agriculture sector in Africa is using data analytics to optimize crop yields, manage resources efficiently, and adapt to climate change. Sensors, drones, and satellite imagery provide valuable data for precision agriculture. A better understanding of the continent's agricultural potential will allow African countries to make better informed decisions on where to invest in agriculture.

– *Space Data:* This has the potential to revolutionize how we understand a wide array of industries and environmental phenomena. African countries are increasingly employing space technologies in addressing several challenges. Satellites communicate by using radio waves to send signals to the antennas on Earth. While getting the satellites' data is imperative, analyzing them properly and effectively knowing how to handle this large volume of data is equally important.

5.7 Challenges

There are several challenges and barriers to BD use in African nations. The African continent has the highest mortality rate in the world, and it is the only region of the world where deaths from communicable diseases exceed deaths from chronic diseases. Reasons for the poor state of healthcare in Africa can be attributed to long-term wars in many parts of Africa, corruption, poor accountability of public office holders, and low spending on health. African cities face significant challenges in achieving the UN SDGs. Africa produces less than 1% of the world's research, even though 12.5% of the world's population is from Africa. As shown in Figure 5.6, there are protection laws in only 33 African nations [27]. Other challenges of BD in Africa include the following [10, 28–30]:

bc Insights

THERE ARE DATA PROTECTION LAWS IN 33 AFRICAN COUNTRIES

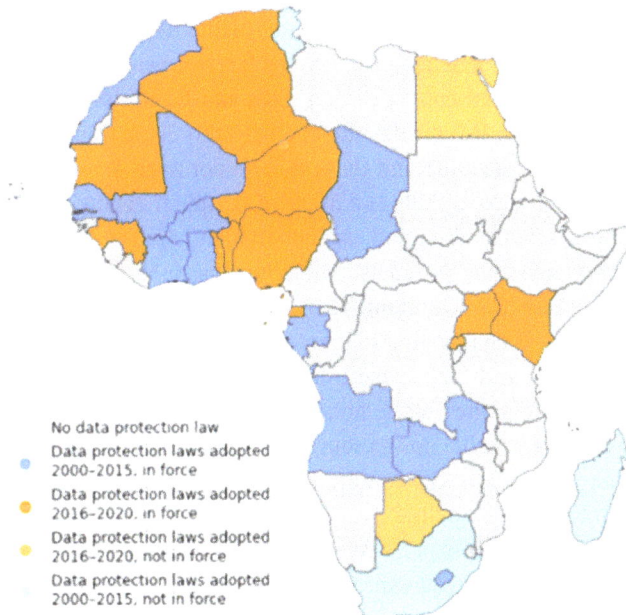

No data protection law

Data protection laws adopted 2000-2015, in force

Data protection laws adopted 2016-2020, in force

Data protection laws adopted 2016-2020, not in force

Data protection laws adopted 2000-2015, not in force

Figure 5.6: There are protection laws in only 33 African nations [27].

- *Ethics:* One major issue is ethics, including the issues of privacy and consent. BD presents some concerns, such as ethical issues in developing African countries, where legal rights and responsibilities are not necessarily fully understood.
- *Environmental Pressure:* The problem of data availability is a major impediment to assessing environmental resilience in the region. Environmental pressures are highest among poorer urban residents living in informal settlements that are exposed to environmental risks such as flooding.
- *Documentation:* Another challenge is the documentation of data accuracies. Documentation may not be easily found in the metadata, or sometimes it may not exist. This affects the accuracy of the resulting mapping and derived products.

- *Privacy:* One significant challenge is how to use healthcare BD in ways that honor individuals' rights to informed consent and privacy. Respecting one's right to informed consent will be very challenging in the era of healthcare BD. Data collected from various sources raises other unique privacy problems. As data is generated in different formats and sources, and across borders using the Internet as a transporting means, individual privacy level becomes an issue.
- *Confidentiality:* Confidentiality concerns the duty to keep disclosed information secret. Clinicians often promise, in the Hippocratic Oath, to respect the fiduciary relationship by protecting confidential information. In a datafied world, it might become challenging to honor confidentiality.
- *Digital Divide:* BD promises to drive economic growth and development, but if not applied across borders, it can lead to a greater digital divide, where nations who successfully embrace BD will advance, leaving the rest behind. The digital divide describes the gap between countries that have and do not have access to computers and the Internet. It indicates the gap between demographics and regions that have access to modern ICT and those that do not have. It does not only refer to physical access but also to skills and usage of computer-related technologies.
- *Workforce:* There is a lack of human capacity in many African nations to use data to develop new policies and products. Data scientists, analysts, and engineers are in short supply across the continent. There are several obstacles on the African continent that make it challenging to create this workforce. For example, power outages are frequent, and high-speed Internet connection is scarce in many places. It is impossible to gather, store, and analyze BD without a dependable Internet connection and a reliable power source. There will be a surge in the demand for skilled data scientists in the coming years to handle all of the technicalities that come with BD and its applications. There are very few institutions across the continent training new BD experts. We need to be prepared for massive job replacement as this technology advances. We need to prepare the youths for high-tech careers.
- *Collaboration:* Governments, corporate entities, and academic institutions need to work together and develop collaborations and partnerships in order to share scarce resources and expertise. We must encourage the development of frameworks that facilitate responsible cross-border data collaboration. This collaborative approach can contribute to a more holistic understanding of BD's role in the digital economy. BD and AI can be used as enablers to foster collaborations among African stakeholders to create a template for a prosperous continent.
- *Gap:* A gap in Africa's data analytics market has been recognized. There is a considerable gap in understanding the data situation in Africa among the users and producers of data. To share experiences and knowledge in data works, a platform to exchange ideas and innovations is critical. SMEs need to be creative in the way they access data that will be useful to businesses on the continent.

- *Brain Drain:* Africa faces a critical challenge of brain drain in its tech sector. The Nigerian phenomenon of "Japa," where young professionals migrate abroad, exemplifies this trend. For example, in 2022 alone, about 500 Nigerian software engineers relocated, predominantly to Canada and European countries, attracted by higher wages.
- *Poverty:* This is a global concern and a major challenge in Africa. Alleviation of poverty is one of the top concerns listed in the SDGs. It is, therefore, imperative for policymakers and other organizations that work toward poverty eradication to have relevant information that will assist them in implementing effective policies and appropriate interventions. Being able to obtain BD on education, health, and standards of living will help in combating poverty in Africa. Satellites used to collect BD can help track poverty in remote areas. Analysis of BD has the potential to provide solutions for those working on poverty eradication.
- *Corruption:* There has been increased corruption in some African nations, and this sometimes leads to resources not being apportioned correctly.
- *More Content Online:* To harness the full potential of BD, there is a need to ensure that we have more content online. Africa needs to be represented more when it comes to data and to tap into this new technology. We need to work on the policy side to digitize our languages. We also need to document and catalog our indigenous knowledge.

These challenges stand in the way of Africa taking advantage of the immense opportunity. African companies and public authorities could still improve the collection and exploitation of BD produced locally in order to take advantage of this huge market if measures are put in place to overcome these challenges. It should be noted that several African governments are working toward addressing these challenges.

5.8 Conclusion

BD is a buzzword that refers to the ability to collect, store, and analyze large amounts of data. It is now an immense opportunity for the continent since it affords Africa a major opportunity to implement solutions that benefit businesses and societies. It offers numerous opportunities for developing economies in Africa to use data analysis in areas such as science, technology, engineering, and mathematics. BD applications have the potential to catalyze the solution of local development challenges in African nations, including applications in key sectors of the economy such as agriculture, healthcare, energy, and resource management. Nonprofits and governments in Africa can also leverage data analytics to improve their services. With the rise of BD, businesses in Africa can gain a competitive advantage over their competitors and develop more effective sales strategies. Unlocking Africa's full economic potential will require a combination of initiatives addressing political and economic challenges as well as building the requisite human capital [23].

In spite of the prevailing challenges and limitations in Africa, there is evidence that BD analytics has the capacity to transform the healthcare system and other industries across the African continent. With the rise of BD, businesses in Africa can gain a competitive advantage over their competitors and develop more effective sales strategies. BD can help African countries improve their governance, reduce corruption, and increase transparency in public services. The future of BD in Africa is frightening but bright. The key is for Africa to ensure it is not left behind. For more information about the use of BD in Africa, one should consult the following related journals:

- *Journal of Big Data*
- *Big Data Research*
- *Big Data & Society*

References

[1] W. Shafik, "Chapter 7 navigating emerging challenges in robotics and artificial intelligence in Africa," https://www.irma-international.org/viewtitle/339985/?isxn=9781668499627

[2] M. N. O. Sadiku, O. D. Olaleye, A. Ajayi-Majebi, and S. M. Musa, "Future of big data," *International Journal of Trend in Research and Development*, vol. 8, no. 5, October 2021, pp. 24–27.

[3] "Introduction to big data analytics," Unknown Source.

[4] M. N. O. Sadiku, O. D. Olaleye, and J. O. Sadiku, "Big data in Africa," *International Journal of Trend in Research and Development*, vol. 11, no. 3, 2024, pp. 132–138.

[5] R. Allen, "Types of big data | Understanding & Interacting with key types (2024)," https://investguiding-com.custommapposter.com/article/types-of-big-data-understandingamp-interacting-with-key-types

[6] "The 42 V's of big data and data science," https://www.kdnuggets.com/2017/04/42-vs-big-data-data-science.html

[7] M. N. O. Sadiku, M. Tembely, and S. M. Musa, "Big data: An introduction for engineers," *Journal of Scientific and Engineering Research*, vol. 3, no. 2, 2016, pp. 106–108.

[8] P. K. D. Pramanik, S. Pal, and M. Mukhopadhyay, "Healthcare big data: A comprehensive overview," in N. Bouchemal (ed.), *Intelligent Systems for Healthcare Management and Delivery*. IGI Global, Chapter 4, 2019, pp. 72–100.

[9] S. Arumugan and R. Bhargavi, "A survey on driving behavior analysis in usage based insurance using big data," *Journal of Big Data*, September 2019.

[10] A. Joubert, M. Murawski, and M. Bick, "Big Data Readiness Index – Africa in the," August 2019, https://link.springer.com/chapter/10.1007/978-3-030-29374-1_9

[11] M. N. O. Sadiku, J. Foreman, and S. M. Musa, "Big data analytics: A primer," *International Journal of Technologies and Management Research*, vol. 5, no. 9, September 2018, pp. 44–49.

[12] C. M. M. Kotteti, M. N. O. Sadiku, and S. M. Musa, "Big data analytics," *Invention Journal of Research Technology in Engineering & Management*, vol. 2, no. 10, Oct. 2018, pp. 2455–3689.

[13] "Future of big data analytics in India," https://datatrained.com/post/future-of-big-data-analytics-in-india/

[14] H. Patel, "What is big data analytics and why it is so important?" May 2018, https://medium.com/@patelharshali136/what-is-big-data-analytics-and-why-it-is-so-important1de86fa37540

[15] P. Ndiho, "AI and big data can drive growth in Africa," https://www.linkedin.com/pulse/ai-big-data-can-drive-growth-africa-paul-ndiho-vvf3e#:~:text=AI%20and%20Big%20Data%20can%20potentially%20transform%20Africa%20and%20drive,of%20the%20thriving%20tech%20industry.

[16] A. Akinnagbe1, K. D. A. Peiris, and O. Akinloye, "Prospects of big data analytics in Africa healthcare system," *Global Journal of Health Science*, vol. 10, no. 6, 2018.

[17] C. Ewuoso, "An African relational approach to healthcare and big data challenges," *Science and Engineering Ethics*, vol. 27, no. 3, May 2021. https://pubmed.ncbi.nlm.nih.gov/34047844/

[18] C. Umezuruike and H. N. Ngugi, "Adoption of big data in educational systems in Sub-Saharan Africa nations," *International Journal of Recent Technology and Engineering*, vol. 8, no. 5, January 2020, pp. 4544–4550.

[19] "Big data in action for government," https://documents1.worldbank.org/curated/en/176511491287380986/pdf/114011-BRI-3-4-2017-11-49-44-WGSBigDataGovernmentFinal.pdf

[20] D. Oni, "Big data and the African space industry," January 2021, https://documents1.worldbank.org/curated/en/176511491287380986/pdf/114011-BRI-3-4-2017-11-49-44-WGSBigDataGovernmentFinal.pdf

[21] T. Hey, "Open science and big data in South Africa," *Frontiers in Research Metrics and Analytics*, vol. 7, November 2022, https://www.frontiersin.org/articles/10.3389/frma.2022.982435/full#:~:text=The%20draft%20South%20African%20Open,used%20to%20analyse%20the%20data.

[23] W. McBain, "Turning data into gold: African analytics accelerates," November 2023, https://african.business/2023/11/technology-information/turning-data-into-gold-african-analytics-accelerates

[24] "Rwanda's progress in embracing big data," https://unstats.un.org/bigdata/regional-hubs/rwanda-concept-note.pdf

[25] J. Bayhack, "Big data and its big potential in Africa," August 2020, https://www.bizcommunity.com/Article/196/662/207385.html

[26] O. Cavine, "Big data analytics for African enterprises," September 2023, https://www.linkedin.com/pulse/big-data-analytics-african-enterprises-osano-cavine

[27] "Data imaginaries and the sub-Saharan African continent," May 2023, https://www.thedatasphere.org/news/data-imaginaries-and-the-sub-saharan-african-continent/#:~:text=African%20people's%20ethos%20of%20interconnectedness,with%20norms%20or%20expectations%20prevailing

[28] E. Banzhaf et al., "Mapping open data and big data to address climate resilience of urban informal settlements in Sub-Saharan Africa," November 2022, https://www.mdpi.com/2225-1154/10/12/186

[29] "ICDL insights: How can Africa address its big data challenges?" April 2023. https://icdl.org/icdl-insights-how-can-africa-address-its-big-data-challenges/#:~:text=To%20do%20this%2C%20it%20is,investing%20in%20infrastructure%20and%20education.

[30] "Using big data to combat poverty in Africa," March 2021, https://furtherafrica.com/2021/03/29/using-big-data-to-combat-poverty-in-africa/

Chapter 6
Cloud Computing in Africa

Cloud is about how you do computing, not where you do computing. – Paul Maritz

6.1 Introduction

The African continent is home to immense potential and abundant resources. It is the home to some of the youngest populations in the world. It is marked by rich diversity, not only in terms of culture, language, and geography but also in the digital land-scapes across the continent. In Africa, a continent known for its dynamic and diverse economic landscape, the adoption of cloud computing (CC) presents both tremendous opportunities and unique challenges.

CC allows individuals and organizations to lease storage and computation resources remotely and as needed. Cloud technology is one of the globally recognized emerging technologies in the new millennium that are most likely to change people's lives. It can bring many benefits to organizations and countries. Organizations with not enough resources to build their own infrastructure can now take advantage of the cloud services to suit their specific needs. Rather than building, owning, and maintaining their own IT infrastructure, businesses can use cloud to access technology resources such as computing capacity, storage, and databases on a pay-as-you-go basis. An industry needs the cloud for the following reasons: (1) Mobile workforce: empowering employees to sift real time data and make decisions on the fly; (2) minimize disruptions: with the right sort of cloud setup problems can be anticipated and solved quickly; (3) collaboration: with the right technology, collaboration – as well as transparency and accountability – are easily managed; (4) innovation: product innovation and process innovation are powerful weapons to survive or thrive in such an environment; and (5) lower cost: no hardware procurement, maintenance, or staff is needed to operate the systems [1].

The African continent is set for significant cloud adoption. Mobility is a major driver of cloud implementations in Africa. Mobile networks in African nations are often more pervasive and robust than wireline connectivity. The majority of African nations such as Algeria, Kenya, Morocco, Nigeria, and South Africa, have either already moved services into cloud hosted infrastructure, or are preparing to do so [2]. Over the last few years, international cloud companies such as IBM, Microsoft, Huawei, Amazon, and Oracle have rushed to develop data centers on the African continent. Apart from these international tech giants, African cloud companies are also making strategic deals across the continent. Data centers, which house digital infrastructure such as servers and cables, enable faster connections and the ability to store the data of tens of millions of people. A typical data center in Malawi is shown

https://doi.org/10.1515/9783112211984-006

in Figure 6.1 [3]. The emergence of local data centers has powered a surge in cloud-based computing in five of Africa's largest economies: South Africa, Nigeria, Kenya, Egypt, and Morocco. Figure 6.2 shows data centers across the continent [4]. There have been significant investments from global and local cloud providers in building African cloud offerings, as well as recent investments in African infrastructure that will support cloud technology, such as fiber-optic cables.

Figure 6.1: A typical data center in Africa [3].

The recent emergence of CC is one of the major advances in the history of computing. CC is a computing paradigm for delivering computing services (such as servers, storage, databases, networking, software, and analytics) over the "the cloud" or Internet with pay-as-you-go pricing. CC has emerged as a transformative force for businesses worldwide, and its imminent impact on Africa is noticeable. Cloud adoption was already a growing trend in Africa for most of the last decade. The African CC market is generating a lot of interest as players position themselves for the boom in data services on the continent. Africa emerges as a landscape of rapid growth and untapped potential. The experience of African countries to date points to CC technology being used at different levels according to the institutions concerned. Cloud-related innovations and services are becoming fundamental sources of economic and societal change in African economies, which are embracing CC [5].

This chapter examines the use of CC in developing African nations. It begins with explaining the concept of CC. It discusses cloud deployment models. It addresses the use of CC in Africa. It covers African nations adapting CC. It highlights the benefits and challenges of big data in Africa. The last section concludes with comments.

Number of commercial data centers* by country (2022 estimate)

< 2   2–4   4–6   6–8   8–20   ≥ 20

15 centers

Morocco

17 centers

Egypt

Ethiopia

Nigeria

Kenya

DR Congo

14 centers

11 centers

South Africa

37 centers

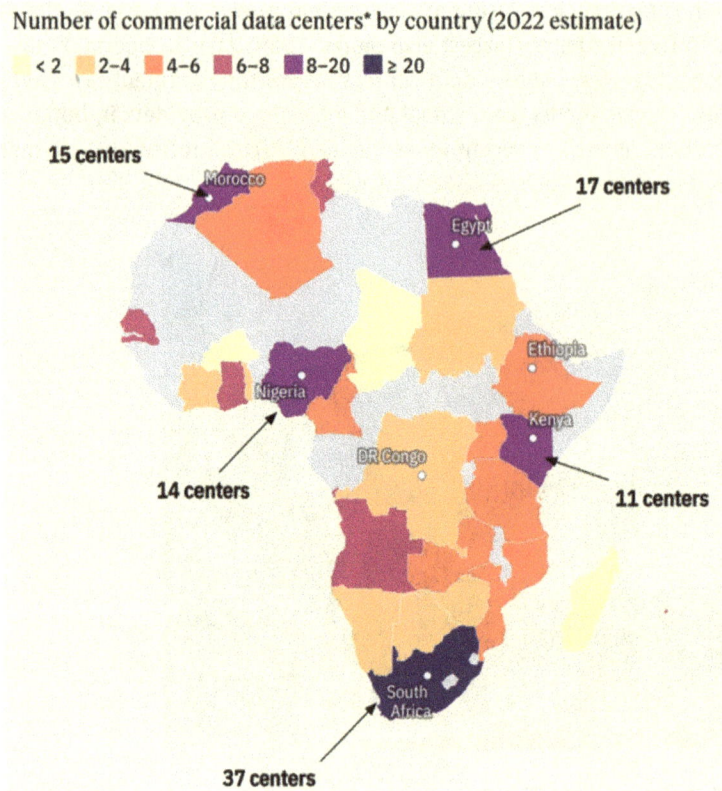

Figure 6.2: Data centers across the continent [4].

6.2 Concept of Cloud Computing

The word "cloud" is a metaphor for describing web as a space where computing has been preinstalled and exist as a service. It originated from the habit of drawing the Internet as a fluffy cloud in network diagrams. The cloud enables you to access your information from anywhere at any time. Figure 6.3 shows a conceptual diagram of CC [6].

CC is a means of pooling and sharing hardware and software resources on a massive scale. Users and businesses can access applications from anywhere in the world at any time. Companies offering these computing services are called cloud providers and typically charge for CC services based on usage. Their aim is making computing a utility such as water, gas, electricity, and telephone services. The main objective of CC is to make a better use of distributed resources and solve large scale computation problems [7].

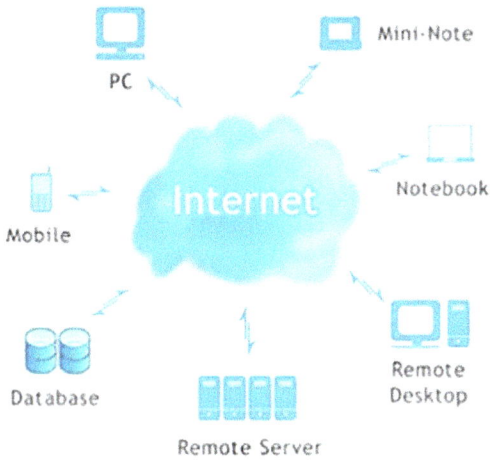

Figure 6.3: A typical cloud computing [6].

CC is not a single produce or piece of technology. Rather, it is a system, primarily providing three different services. The services provided by CC are shown in Figure 6.4 and explained as follows.

- *Infrastructure-as-a-Service* (IaaS)*:* This is the simplest of CC offerings. It involves the delivery of huge computing resources such as the capacity of storage, processing, operating systems, servers, computing power, firewalls, bandwidth, and network which form the underlying cloud infrastructure. It allows users to rent any form of hardware and software, and remotely access computing resources on a pay-per-use basis. The major advantages of IaaS are pay per use, security, and reliability. IaaS is also known as hardware-as-a-service (HaaS). An example of IaaS is the Amazon Elastic Compute Cloud, IBM and Amazon use IaaS.
- *Platform-as-a-Service* (PaaS)*:* This supports the development of web applications quickly and easily. The customer can build his own applications, which run on the cloud provider's infrastructure. It has emerged due to the suboptimal nature of IaaS for CC and the creation of web applications. Many big companies seek to dominate the platform of CC, as Microsoft dominated personal computer (PC). Examples of PaaS are Google App Engine and Microsoft Azure.
- *Software-as-a-Service* (SaaS)*:* This is also known as software-as-a-server or software-on-demand. This provides a service (software applications over the Internet) that is directly consumable by the end-user. It is software deployed over the Internet. This is a pay-as-you-go service. It seeks to replace the applications running on PC. Google, Twitter, and Facebook are typical examples of SaaS. Gmail, AOL, Yahoo, and Skype all use SaaS.

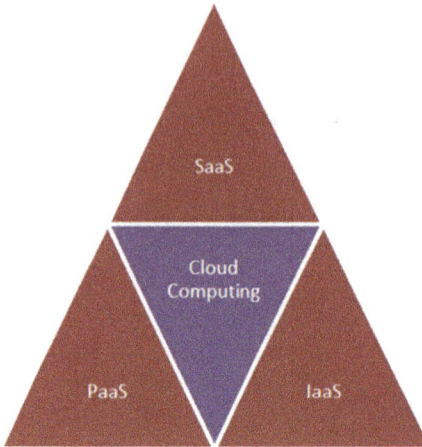

Figure 6.4: Cloud computing services.

As CC becomes mature, several service types are being introduced and overlaid on these architectures. These include HaaS, application-as-a-service, network-as-a-service, data-storage-as-a-service, IT-as-a-service, functions-as-a-service, cooperation-as-a-service, and municipality-as-a-service. Figure 6.5 illustrates some features of CC [8].

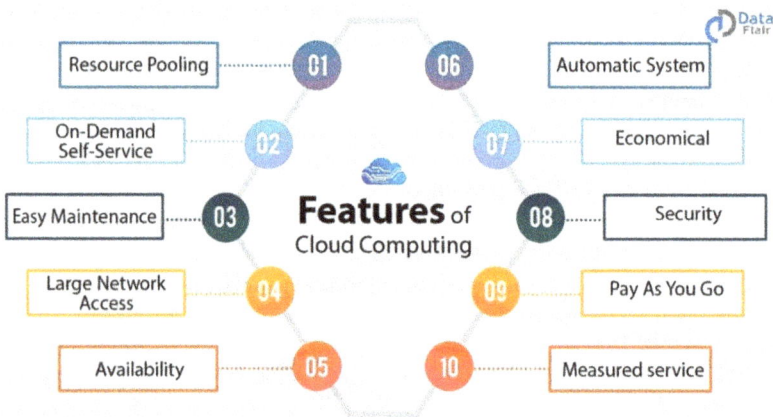

Figure 6.5: Some features of cloud computing [8].

The major vendors of CC include IBM, Microsoft, Cisco, Oracle, Sun, and Siemens. Several companies including IBM, Microsoft, Google, Verizon, Amazon, Rackspace, Eucalyptus, and Netflix have started to offer diverse cloud services to their customers. These are the dominant CC products. Others include Force.com from Salesforce, Dropbox, iCloud (Apple) used globally, Ucloud (KT) emerging in Korea, Mobile office (KT), and Mobile share (AT&T). Charges for cloud services are based on three key items: storage, bandwidth, and compute. Storage is the amount of data (in GB) stored over a monthly period. Bandwidth is calculated at the amount of data (in GB) transferred in and out of platform service through batch processing and transaction.

6.3 Cloud Deployment Models

Application and databases are moved to the centralized data centers, known as clouds. There are three deployment models of cloud. Figure 6.6 shows their relationship [9]:

–   *Public Cloud:* This describes CC in the traditional mainstream sense. It is characterized by public availability of the cloud services. It allows users to access the cloud through interfaces using web browsers. It is when the services are provided over a network that is open for the public. It is a cost effective, elastic means of deploying a solution. It is typically based on a pay-per-use model. It is less secure than the other cloud models. Typical examples of public cloud are Google and Facebook. Public cloud service providers operate the infrastructure at their data center and provide access through the Internet.
–   *Hybrid Cloud:* This is the cloud environment that offers the benefits of multiple deployment models. It provides solutions through a mix of both private and public clouds. Its main strength is that it provides better scale and convenience of a public cloud as well as the control and reliability of private CC. Its adoption depends on several factors such as data security, compliance requirements, and the level of control needed over data.
–   *Private Cloud:* This describes offering that makes cloud services available for a designated single and private organization. It is created within an organization's data center. It is different from public cloud in that all the resources and applications are managed by the organization. It is more secure than the public cloud because only the organization stakeholders are allowed to access the private cloud. When appropriately constructed, it can improve business. Eucalyptus was the first to deploy private clouds.

Apart from these, other CC deployment models include community CC, distributed cloud, shared private CC, dedicated private CC, multicloud, and dynamic private clouds.

6.4 Use of Cloud Computing in Africa

CC is viewed as having the potential to significantly bolster economic growth through the provision of cost savings and efficiencies. Enterprises of all sizes and across sec-

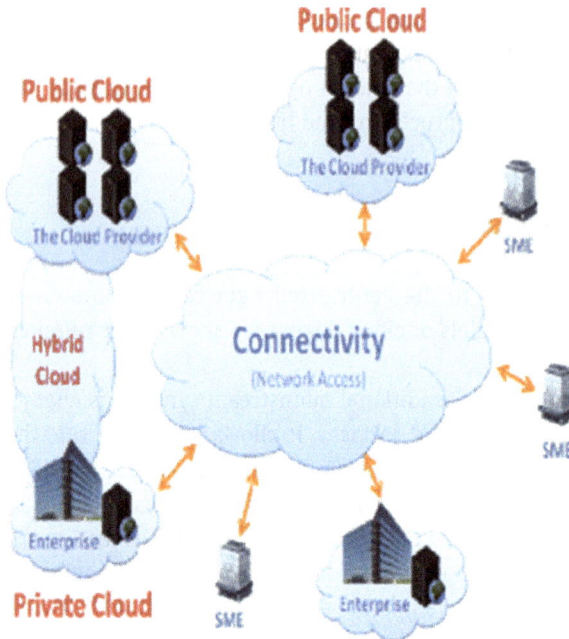

Figure 6.6: Deployment model types [9].

tors are steadily embracing CC. CC has an impact on a variety of industries, including education, healthcare, banking, manufacturing, finance, agriculture, government, and communication. African banks, insurance companies, airlines, and airports are moving their data and IT systems to virtual servers and shuttering costly self-operated data centers. We consider the following areas where CC has made the greatest impact in Africa [2]:

- *E-education:* This is also referred to as online learning. It is learning facilitated any type of education that is enhanced by the use of information and communications technology (ICT) and online communication. It allows students to access a central curriculum and knowledge repositories based in the cloud that allow users to take courses electronically. E-education has led to M-Learning, or mobile learning, the use of mobile phones as the primary education tool. While e-education is not the magic bullet for a nation's education woes, it could greatly improve the system with less investment required.
- *E-health:* This is a serious focus area in Africa because of the medical challenges faced by the continent, with HIV/AIDS, TB, and Malaria infesting parts of the continent. It is the use of information and communication technologies for health systems. There is lack of doctors and medical facilities is in Africa, making e-health an obvious benefit. CC in healthcare has facilitated the rise of telemedicine platforms, connecting patients in remote areas with healthcare professionals.

- *E-commerce:* This is commerce conducted electronically or on the Internet. Buying and selling, online banking, supply chain management, and teleconferencing can all fall under e-commerce. Financial services organizations had the lowest rates of cloud adoption. E-commerce mobile penetration is going to be a good indicator for the success of a solution like the M-PESA to take hold. As an e-commerce trailblazer in Africa, Jumia's use of cloud technology has been pivotal in scaling its operations across various African nations, ensuring efficiency and reliability in its online marketplace.
- *Financial Services:* Digital financial services providers and other technology start-ups become an embedded part of African economies. African banks have already realized the value of cloud banking. Before financial service providers can adopt cloud banking, regulators need to support and approve the use of cloud technology within the financial sector
- *Government:* Governments are hesitant to move data to public cloud platforms. On-premise data systems or contracts with smaller cloud service providers continue to dominate government cloud spending. Developing supportive government policies and regulatory frameworks can accelerate cloud adoption and foster a conducive environment for technological growth.
- *Agriculture:* The backbone of many economies in Africa is the agricultural sector. Leveraging cloud-based IoT and data analytics, African farmers are improving crop yields and optimizing resource use, leading to more sustainable and productive agricultural practices.
- *Insurance:* The insurance sector is leveraging CC for risk assessment, policy management, and customer engagement, transforming traditional models into agile, data-driven enterprises.

6.5 Adapting Cloud Computing in African Nations

A boom in the digital economy in Africa is boosting demand for online data storage by banks and mobile telecommunication companies across the continent. In September 2021, IBM secured cloud deals with some of Africa's largest banks in Morocco, South Africa, Nigeria, and Mozambique. Oracle recently celebrated the opening of new offices in Nigeria, Kenya, Ghana, Ivory Coast, Mauritius, South Africa, Egypt, Morocco, and Algeria. CC optimizes the use of scarce resources by consolidating what is available into a resource pool and is set for a major take off in Africa, provided certain enabling measures are put in place. We consider the cloud provision in some selected African nations:

– *South Africa:* The fourth industrial revolution, which is powered by cloud-led technologies, has significantly accelerated in South Africa. In spite of its floundering power infrastructure, South Africa remains the favorite destination for data centers. Microsoft Amazon, Google, Huawei, and Oracle have all built their data centers in

South Africa. Due to high concentration of data centers in South Africa, the country often prefers owing to its relatively better infrastructure and opportunities for synergies. South African Airways transitioned to CC, shedding its servers to migrate to the cloud. CC model is already being adopted as a way of overcoming compute capacity in South Africa as the leading industry for now. The Oracle Cloud Johannesburg Region will boost cloud adoption across Africa while also helping businesses achieve better performance and drive continuous innovation. It offers a next-generation cloud to run any application faster and more securely for less, helping businesses build resilience, agility, and achieve improved ROI. Partnership with Oracle has always been one of Accenture's most strategic and important initiatives to help our clients leverage the cloud and thrive in a cloud-first world [10]. The availability of Amazon Web Services, Microsoft Azure, Oracle, Huawei, Google, and Oracle cloud data centers in South Africa provides companies with the means to grow and expand in a digital environment. The importance of these cloud data centers in South Africa cannot be overstated [11].

– *Nigeria:* In Nigeria, state governments are reducing levies on the development of telecom infrastructure. When laying fiber cables, state governments demanded the Right of Way (RoW) levy for every meter of fiber being deployed. RoW taxes influence how and where telecom infrastructure operators decide to lay cables. This has an impact on mobile penetration. Companies in the finance, and oil and gas sectors that traditionally maintain in-house servers have switched to the cloud [12]. It is worth noting that while African companies are progressively transitioning to the cloud, a majority still rely on foreign service providers with data centers situated overseas. This reliance is notably seen in Nigerian government agencies, with 70% of them choosing to host their data abroad, thus limiting the ability of the data center operators to offer the fullest range of services. Factors such as cost, reliability, and data storage size often lead African firms and governments to prefer offshore hosting. When a central bank demonetization program forced Nigerians to use digital money transfer options, the money transfer services offered by Nigerian banks were frequently down and failed transactions were a common complaint. By harnessing CC, Paystack transformed online payment processing in Nigeria. Through transition from telco to techco, they can now provide a portfolio of products and services that grow the digital economy in Nigeria.

– *Kenya:* Kenya has ambitious plans to diversify its economies through digitization. It has distinguished itself as a hotbed of technological innovation and creativity. It is currently the leader in Africa when it comes to Internet and mobile penetration. It has one of the highest mobile subscriptions in Africa and the highest share of Internet usage from mobile phones compared to desktops. It has implemented several data protection regulations, including the Data Protection Act 2019, which aims to protect personal data from unauthorized access and use. In 2022, the Kenyan government initiated a 10-year Digital Masterplan 2022–2032 for leveraging and deepening ICT contribution toward accelerating economic growth. Oracle is planning to open a new cloud region in Kenya to accelerate the digital transformation of its government and private sector. This is an effort to meet the growing demand for Oracle Cloud Infrastructure services across Africa. Oracle traditionally relies on data center partners rather than building its own facilities [13, 14]. Angani, a Kenyan startup, is seeking to become the market leader not just locally but in the region with provision of cloud solutions.

– *Somali:* Along with increased focus on infrastructure, Somali Telecom Industry have made major investments in the ICT landscape and presents a large, growing market that is increasingly connected via mobile technology and broadband connectivity. This is critical, as a lack of effective communications infrastructure has traditionally been one of the biggest obstacles to economic growth. Somalia is poised for the next wave of technology innovation where IT services will be instantly available to end users on request [15].

6.6 Benefits

CC is a new computing paradigm that can provide remote access to these resources that were otherwise inaccessible. As in all parts of the world, CC brings unquestionable benefits to the ICT sector. It allows startups companies achieve real productivity by giving them access to the necessary tools at low costs. With CC, businesses and individuals can access these resources on demand, paying only for what they use. It enables users to access a wide range of services and applications from anywhere in the world, as long as they have an Internet connection. Africa is emerging as a critical player in the global cloud ecosystem. Figure 6.7 shows a data technician, demonstrating how Africa is ready [16]. Other benefits of CC in Africa include the following [17, 18]:

- *Cost Savings:* The most compelling reason to move to cloud is undoubtedly cost savings. One of the benefits of cloud technology is cutting costs to save funds. In multiple ways, a cloud based call center sets up on the cloud benefit African businesses to reduce expenses. Cloud technology makes everything remote and hosted on the internet, which saves a lot of money for these businesses. One does not need to put staff to manage IT tools and software or servers. Cloud technology platforms cover basic maintenance and support, which reduces the cost of maintenance and support for the contact center industry of Africa.
- *Digital Transformation:* Cloud transformation is the process of migrating work to the cloud, which can include data, apps, software programs, etc. CC is a catalyst for complete digital and cloud transformation, enabling businesses to modernize their operations.
- *Innovation:* One of the most compelling benefits of CC is its ability to provide businesses in Africa with access to state-of-the-art technologies. CC allows even small

and medium-sized enterprises to harness powerful computing capabilities without the need for extensive hardware. This democratization of technology has the potential to drive unprecedented levels of innovation and competitiveness across industries. Cloud services provide the agility necessary to respond swiftly to changing market demands

- *Growth:* Companies in Africa have shown they are no laggards as far as adopting cloud is concerned. As of 2023, 50% of African companies have already adopted cloud capabilities in all or most parts of their business and there are no signs of slowing down.
- *Digital Natives:* Africa's digital natives will power the future of the Internet. They will naturally adopt Internet services for everything from education to payment, and represent new possibilities for the digital economy in Africa. Some digital natives are shown in Figure 6.8 [19].
- *Customer Services:* Cloud technology is flexible and supports remote operations. It can access all tools, software, platforms, and data from remote locations. Thus, customers can be attended to anytime as the software will be available to use for agents to cater to clients. Even contact centers based in Africa can hire customer service experts from any nook and corner of the world to level up the game of customer care.
- *Scalability:* The role of cloud technology in the contact center industry of Africa can be defined by its capacity to scale up the software and infrastructure. Cloud technology in the African call center industry is leading this area by providing instant scalability options. The IT infrastructure of the company has to be compatible enough and scalable enough to support the growth of businesses operating in the contact center industry of Africa.
- *Efficiency:* Cloud technology helps agents in Africa to work from anywhere. This helps in improving the efficiency of their work by letting them enjoy work life balance. This boosts their morale and contributes to elevating the efficiency of agents.

6.7 Challenges

Challenges abound, however, in capturing this value. The main challenges are the need for adequate infrastructure, limited bandwidth provisions, point of presence issues, latency issues, and workload complexities in the cloud, wide variations in language, culture, and currency exist, differing regulatory environments, limited infrastructure, and regulatory policies. Other challenges of CC in Africa include the following [15, 20–24]:

- *Limited Connectivity:* Foreign investors are keen on the African CC market, which currently has a penetration rate of about 15%. While the promise of CC is vast, the

Figure 6.7: A data technician, showing Africa is ready [16].

Figure 6.8: Some digital natives [19].

reality of Internet accessibility in Africa presents a significant challenge. Connectivity remains uneven across the continent, with some regions still grappling with limited or unreliable access to the Internet.

- *Digital Divide:* The digital divide in Africa refers to the gap between those who have access to digital technologies and the Internet, and those who do not. This necessitates concerted efforts from governments, private sector stakeholders, and international organizations to invest in infrastructure and expand broadband access, ensuring that all businesses can fully harness the potential of the cloud.
- *Data Security and Privacy:* In an era marked by increasing cyber threats, safeguarding data is paramount. African businesses must navigate the complex landscape of data security and privacy regulations. Cloud service providers must establish robust security measures, including encryption, access controls, and regular security audits to instill confidence in users. Data protection laws and privacy regulations vary from nation to nation, adding a layer of complexity to

cloud adoption. Robust security measures are essential for safeguarding sensitive information.

- *Regulation:* Navigating the regulatory landscape is crucial for businesses operating in Africa. Data residency laws and data protection laws are common across Africa and require regulated data (such as personal information) to be localized within the country's borders. Given the limited data center presence of cloud providers in Africa, this regulation essentially makes it impossible for many organizations to use public cloud for these data sets. In the future, regulatory bodies, organizations, and service providers will need to collaborate to establish a harmonized regulatory framework.
- *Dominance:* There are concerns about the growing dependence by companies on the cloud as it is dominated by just a few tech giants. In the banking industry, for example, the oligopolistic global CC market is the perfect recipe for the ultimate archetypal centralization.
- *Infrastructure:* Investments in Internet infrastructure and stable power supply are crucial. Strengthening Internet infrastructure is vital for widespread cloud adoption. The cost of increasing capabilities that often require new infrastructure is often prohibitively expensive for small companies. Collaborative efforts between governments and private entities can accelerate this development. There is the fear that without this much needed infrastructure; much of the continent could be left behind.
- *Power:* Basic infrastructure like electricity remains in short supply in many African countries. The running of data centers is power intensive. That presents a big hurdle in a continent that is struggling to bridge a large power deficit. Data centers, therefore, cannot depend on grid power. The power infrastructure constraint impedes the development of the foundations upon which faster Internet infrastructure must be laid on. Electricity via diesel-powered generators is expensive and limiting.
- *Data Colonialization:* Africa's data has traditionally been stored in other parts of the world, through a network of undersea cables, causing latency. Africa is highly vulnerable to data colonialization. Africa's over-dependence on the US for cloud services is now a breeding ground for data colonialism, where developed economies prey on Africa's large-scale data resources from which they extract economic value.
- *Cloud Readiness:* The state of "cloud readiness" is determined by the knowledge level of how technology, services, and solutions will take advantage of expertise and how best to implement cloud technology. This readiness will be examined by comparing socioeconomic and wireless connectivity indicators on the African continent. These indicators were chosen to show the push and pull factors of movement to the cloud.
- *Job Loss:* There is a level of ignorance in the African market where the impact of CC on jobs is concerned. Many believe CC means moving your IT services to an

overseas public cloud infrastructure, resulting in job losses. On the contrary, cloud enables new levels of business agility that creates new job opportunities.

- *Skills Gap:* The development of CC skills is critical, if Africa is to benefit from CC. CC presents an opportunity for professionals seeking to be relevant in the years to come. In South Africa, Oracle has initiated an annual "Graduate Leadership Program" to address the skills gap in the local ICT industry and creating employment opportunities for South Africa's youths. The skills young Africans need in this age include [17, 23]: (1) blockchain, (2) CC, (3) analytical reasoning, (4) artificial intelligence, (5) user experience/design, (6) graphics design, (7) communication, (8) people management, and (9) coding.

By addressing these challenges, African businesses and economies stand at the threshold of a major digital transformation and Africa will emerge as a significant player in the global digital landscape. The solution for Africa's challenges will come from within Africa. While the road to widespread cloud adoption in Africa presents its unique set of challenges, the opportunities far outweigh these hurdles.

6.8 Conclusion

CC enables companies to consume IT resources (such as hardware, software, and storage) as a utility, just like electricity, rather than having to build and maintain computing infrastructures in-house. It has allowed users all over the globe to undertake mission critical projects from different part of the world. With cloud transformation enabling businesses to navigate challenges, Africa's cloud-driven future holds great promise. The urgent case for cloud transformation in Africa is clear. African nations and organizations that embrace cloud technologies strategically and swiftly will be better positioned to unlock value and drive innovation [25].

The use of cloud services in Africa is on the rise. While Africa represents an opportunity for cloud service providers, the market is still small. With cloud infrastructure, African businesses can compete on a global stage, offering services and products beyond local markets. Data tomorrow is what oil is today and the battle is on who will control it. For more information about the use of CC in Africa, one should check the following related journals:

- *Journal of Cloud Computing*
- *IEEE Cloud Computing*
- *IEEE Transactions on Cloud Computing*
- *International Journal of Cloud Applications and Computing*
- *International Journal of Cloud Computing and Services Science*
- *i-manager's Journal on Cloud Computing*

References

[1]   W. Shafik, "Chapter 7 navigating emerging challenges in robotics and artificial intelligence in Africa," https://www.irma-international.org/viewtitle/339985/?isxn=9781668499627

[2]   C. E. O. Otieno, "Benefits of cloud computing in Africa," March 2016, https://www.linkedin.com/pulse/benefits-cloud-computing-africa-charles-evans-ogego-otieno

[3]   A. Akwagyiram, "New data centers are supercharging cloud computing in smaller African countries," https://www.semafor.com/article/06/22/2023/data-centers-fuel-cloud-computing-in-smaller-african-countries

[4]   A. Akwagyiram, "New data centers are supercharging cloud computing in smaller African countries," https://www.yahoo.com/news/data-centers-supercharging-cloud-computing-172113928.html

[5]   M. N. O. Sadiku, O. D. Olaleye, and J. O. Sadiku, "Cloud Computing in Africa," *International Journal of Trend in Research and Development*, vol. 11, no. 3, 2024, pp. 125–131.

[6]   C. Nkhata, "A simple explanation of 'cloud computing' for non-techie people," http://www.chenkankhata.com/2011/06/simple-explanation-of-cloud-computing.html

[7]   M. N. O. Sadiku, S. M. Musa, and O. D. Momoh, "Cloud Computing: Opportunities and Challenges," *IEEE Potentials*, Jan./Feb. 2014, pp. 34–36.

[8]   "Features of cloud computing – 10 major characteristics of cloud computing," https://data-flair.training/blogs/features-of-cloud-computing/

[9]   B. Grobauer, T. Walloschek, and E. Stocker, "Understanding Cloud Computing Vulnerabilities," *IEEE Security and Privacy*, vol. 9, no. 2, March/April 2011, pp. 50–57.

[10]   "Oracle opens first cloud region in Africa," January 2022, https://www.oracle.com/news/announcement/oracle-cloud-johannesburg-region-2022-01-19/

[11]   "South African businesses are ready for the cloud," February 2023, https://www.instrumentation.co.za/18334r

[12]   A. Idris, "Africa's cloud computing industry is set to grow as data adoption rises," May 2020, https://techcabal.com/2020/05/29/africas-cloud-computing-industry-is-set-to-grow-as-data-adoption-rises/

[13]   C. M. Nana, "Cloud computing can improve local realities in Kenya," April 2018, https://cioafrica.co/cloud-computing-can-improve-local-realities-in-kenya/

[14]   D. Swinhoe, "Oracle plans cloud region in Nairobi, Kenya," February 2024, https://www.datacenterdynamics.com/en/news/oracle-plans-cloud-region-in-nairobi-kenya/#:~:text=Oracle%20opened%20its%20Johannesburg%20OCI,is%20planned%20in%20the%20future.

[15]   I. D. Osman, "Somali federal government's cloud initiative: challenges and promise," February 2015, https://hiiraan.com/op4/2015/feb/98253/somali_federal_government_s_cloud_initiative_challenges_and_promise.aspx#:~:text=While%20cloud%20holds%20a%20lot,sensitive%20and%20personal%20identifiable%20information.

[16]   "Is Africa ready for the cloud computing revolution?" November 2023 https://www.telecomreviewafrica.com/en/articles/features/3922-is-africa-ready-for-the-cloud-computing-revolution

[17]   "The urgent case for cloud transformation in Africa https://theexchange.africa/tech-business/cloud-transformation-africa/

[18]   "The role of cloud technology in the growth of Africa's contact center industry," April 26, 2023, Unknown Source.

[19]   "7 skills every young African should have," May 2016, https://www.weforum.org/agenda/2016/05/7-skills-every-young-african-should-have/

[20]   S. Blumberg, J. Gelle, and I. Tamburro, "Africa's leap ahead into cloud: Opportunities and barriers," January 2024, https://www.mckinsey.com/capabilities/mckinsey-digital/our-insights/africas-leap-ahead-into-cloud-opportunities-and-barriers

[21]   "Is Africa ready for the cloud computing revolution?" November 2023, https://www.telecomrevie
        wafrica.com/en/articles/features/3922-is-africa-ready-for-the-cloud-computing-revolution

[22]   R. Raji, "Where is Africa in the cloud? https://www.ntu.edu.sg/cas/news-events/news/details/where-
        is-africa-in-the-cloud

[23]   D. Govender, "How cloud computing is revolutionising African businesses," February 2024,
        https://www.sovtech.com/blog/how-cloud-computing-is-revolutionising-businesses#:~:text=Global%
        20Market%20Access%3A%20With%20cloud,decision%2Dmaking%20and%20strategic%20planning.

[24]   F. Ngila, "The scramble for Africa's data is taking place on the cloud," August 2022, https://qz.com/
        the-scramble-for-africas-data-is-taking-place-on-the-cl-1849444808#:~:text=The%20demand%20for
        %20cloud%2Dbased,secure%20generates%20huge%20extra%20costs.

[25]   "Six tech skills African youngsters need in this day and age," https://thellpafrica.org/six-tech-skills-
        youngsters-need-in-this-day-and-age/

Chapter 7
Internet of Things in Africa

Anything that can be connected will be connected – Morgan.

7.1 Introduction

Africa is a booming continent with incredible growth potential and is the second-largest continent in the world. It is a continent that has 54 countries, with an area of 30,370,000 km^2 and 1.4 billion individuals as of 2021, subdivided into 5 major regions, like Northern Africa (with countries like Libya, Egypt, North Sudan, Algeria, Morocco, and Tunisia, as demonstrated) inhabiting the northerly region of Africa [1]. The continent is not just catching up with the world; it is propelling itself to the forefront of innovation. Africa is rising, and its tech scene is leading the way. Africa is closely watched as the next big growth market. It is home to some of the youngest populations in the world. Africa is a booming continent with incredible growth potential, as the second-largest continent in the world and the world's largest free trade area, connecting 1.3 billion people (16.6% of the world population) across 55 nations.

Today, the Internet has become an indispensable part of life. When it comes to the Internet, the Internet of things (IoT) has taken center stage. The IoT is a giant network of connected things and people. The idea behind creating IoT was the amalgamation of the physical world into computer-based systems. Some examples of IoT devices are cell phones, washing machines, laptops, etc. With IoT, anything that can be connected shall be connected. This is best illustrated in Figure 7.1 [2]. The IoT is a tool for connecting physical objects to the virtual world using sensors and some Internet protocols to lessen human interventions.

The IoT refers to the billions of physical devices connected to the wireless Internet that allows exchanging data around the world. It is one of the disruptive technologies and is growing rapidly. It is an integral part of the Future Internet, where physical and virtual "things" have identities and are seamlessly integrated into the information network. With IoT technology, we are fast moving toward a society where everything and everyone will be connected. IoT is transforming the education sector and making learning simpler and faster [3–5].

From financial services to agriculture, digital technology is being leveraged to deliver greater access and usher in the "future of everything" in Africa. Figure 7.2 shows smart Africa's vision statement [6]. The African continent has been slower in embracing the IoT concept compared to most developed nations, but Africa is now increasing its level of intake of IoT. The rise of IoT in Africa is driven by several elements, such as a need to digitize industries such as manufacturing, transportation, agriculture,

https://doi.org/10.1515/9783112211984-007

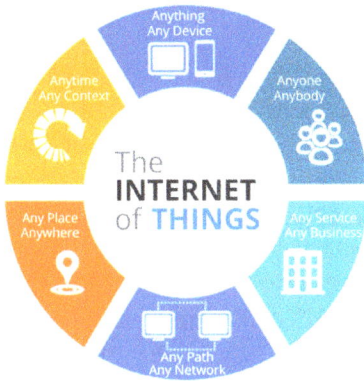

Figure 7.1: Anything that can be connected is connected to IoT [2].

and retail. The African continent as a whole is also seeing an increasing adoption of IoT. Several African nations, such as South Africa, Kenya, and Nigeria, are investing in IoT to improve their infrastructure, increase efficiency, and reduce costs [7].

Figure 7.2: Smart Africa's vision statement [6].

The IoT is the next great revolution in global industry, technology, and life. It is one technology that is penetrating the world so fast and is being adopted to create smart homes, smart environments, smart cities, connected automobiles, wearables, and the industrial Internet. Companies worldwide use IoT technology to improve efficiency and accessibility in their everyday operations. IoT is recently emerging for African enterprises that are innovating and establishing technological growth. The IoT is gaining momentum across Africa and has the potential to solve many problems on the continent. It is a real chance for Africa to be part of the global economy. The IoT market in Africa is expected to experience significant growth in the coming years.

This chapter examines the adoption of IoT in African nations. It begins with explaining the concept of the IoT. It addresses the use of IoT in Africa. It covers some African nations adapting IoT. It highlights the benefits and challenges of IoT in Africa. The last section concludes with comments.

7.2 Concept of Internet of Things

The term "Internet of things" was introduced by Kevin Ashton from the United King-
dom in 1999. IoT is a network of connecting devices embedded with sensors. It is a
collection of identifiable things with the ability to communicate over wired or wireless networks. The devices or things can be connected to the Internet through three
main technology components: physical devices and sensors (connected things), connection and infrastructure, and analytics and applications.

The IoT is a worldwide network that connects devices to the Internet and to each
other using wireless technology. IoT is expanding rapidly, and it has been estimated
that 50 billion devices will be connected to the Internet by 2020. These include smart
phones, tablets, desktop computers, autonomous vehicles, refrigerators, toasters, thermostats, cameras, pet monitors, alarm systems, home appliances, insulin pumps, industrial machines, intelligent wheelchairs, wireless sensors, and mobile robots. A typical IoT is shown in Figure 7.3 [8].

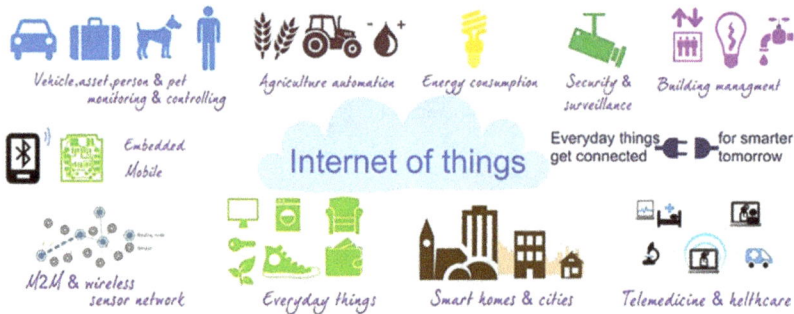

Figure 7.3: A typical IoT [8].

The IoT is much more than a simple technology. There are four main technologies
that enable IoT [9]:
1. Radio-frequency identification (RFID) and near-field communication.
2. Optical tags and quick response codes: These are used for low-cost tagging.
3. Bluetooth low energy.
4. Wireless sensor network: They are usually connected as wireless sensor networks
   to monitor physical properties in specific environments.

Other related technologies are cloud computing, machine learning, and big data.

The IoT technology enables people and objects to interact with each other. It is
employed in many areas, such as smart transportation, smart cities, smart energy,
emergency services, healthcare, data security, industrial control, logistics, retails, government, traffic congestion, manufacturing, industry, security, agriculture, environ-

ment, and waste management. Figure 7.4 shows the most widely used application areas of IoT [10].

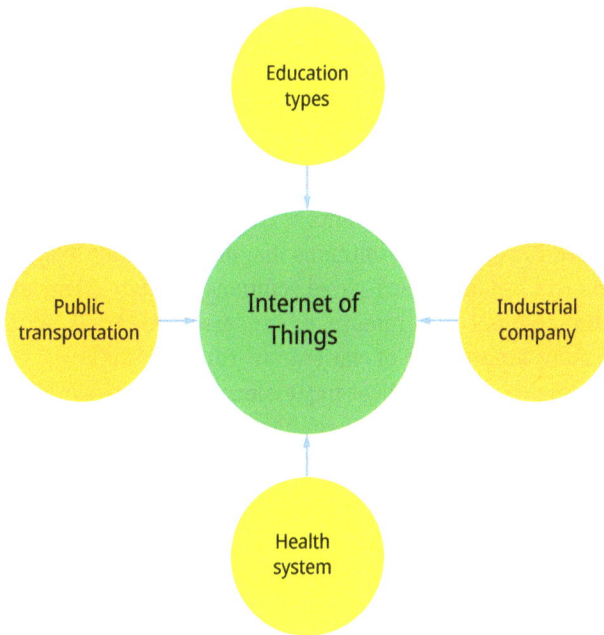

Figure 7.4: The most widely used IoT application areas [10].

IoT supports many input-output devices such as cameras, microphones, keyboards, speakers, displays, microcontrollers, and transceivers. It is the most promising trend in the healthcare industry. This rapidly proliferating collection of Internet-connected devices, including wearables, implants, skin sensors, smart scales, smart bandages, and home monitoring tools, has the potential to connect patients and their providers in a unique way.

Today, the smartphone acts as the main driver of IoT. Smartphones are equipped with healthcare applications.

The narrowband version of IoT is known as narrowband IoT (NBIoT). This is an attractive technology for many sectors, including healthcare, because it has been standardized [11]. The main feature of NBIoT is that it can be easily deployed within the current cellular infrastructure with a software upgrade.

The IoT is essentially the connection of devices to the Internet.

IoT network connects various types of devices, like tablets, smartphones, personal computers, laptops, and wearable devices.

7.3 Use of Internet of Things in Africa

The IoT has revolutionized the way we connect with and interact with our environment. IoT technology significantly impacts the lives of people across Africa. It is already making its presence felt on the African continent by addressing a host of issues around security, water control, energy, health monitoring, mining, agriculture, manufacturing, government, and traffic congestion. Here, we consider IoT in some selected industries:

– *Healthcare:* Integrating IoT-based solutions in healthcare transforms healthcare delivery in Africa. IoT is being used for healthcare solutions that provide better diagnoses and remote treatment options for patients. IoT-based solutions facilitate telemedicine consultations, connecting patients in remote or underserved areas with healthcare professionals. Remote monitoring of patients in rural or underserved areas is now possible, connecting them to medical professionals miles away. RFID tags track medical equipment and supplies in the healthcare sector. Hospitals and clinics can better manage resources, reduce waste, and improve patient care.

– *Automotive Industry:* The automotive industry is using sensors and beacons embedded in the road, working together with car-based sensors, to be used for hands-free driving, traffic optimization, and accident avoidance. Automotive IoT covers IoT use cases in mobility and transportation settings. Among the various segments within the IoT market, the automotive IoT sector is expected to dominate with a projected market volume of US$9.17 bn in 2024. This indicates the increasing integration of IoT technology within the automotive industry [12]. An IoT environment delivers more intelligence on the assembly line or in the manufacturing process.

– *Agriculture:* Agriculture is a key player in nearly all African nations, and it occupies a pivotal role in the improvement of the continent's economy. Digital agriculture is an ever-expanding field focused on the enhancement of farming through improved information and communication processes. An IoT sensing platform can make digital farming accessible to small-scale farmers in rural Africa. It helps in enhancing agriculture productivity by manifolds. The platform provides information on the state of the soil and surrounding environment, coupled with computer vision to classify the type of soil. It has been tested and deployed in real work environments in Uganda and South Africa [13]. Farmers can attach RFID tags to animals, and this helps in disease control, breeding management, and tracking livestock movements. For example, agricultural land in Nigeria was found to be 77.7% in 2016, while employment in agriculture was 36.55% in 2017. This makes agriculture an important sector in Nigeria.

– *Mobile Applications*: In Africa, mobile numbers are people's unique identifiers for digital services. IoT mobile applications play a significant role in connecting people and data by enabling devices to communicate with each other and users through smartphones or tablets. They empower individuals to harness the potential of IoT technology to improve efficiency, convenience, and security. These applications can analyze data from connected devices to provide insights and trends, helping users make informed decisions. IoT mobile apps enable users to access and control their IoT devices from anywhere with an Internet connection, increasing convenience and accessibility [14].

– *Business:* African businesses are grappling with the new reality of the digital age. At its core, digital transformation fundamentally changes everything for businesses. With the massive number of interconnected things, businesses all over the world are positioning themselves to tap into the huge potential that IoT brings. Businesses in countries across Africa are now using IoT applications to improve their business environment as well as the lives of citizens.

– *Government:* IoT technology is being used by government agencies and commercial companies to make vital services, such as utilities, more efficient and accessible. Reforming governments have started to appear across the length and breadth of Africa.

– *Entertainment:* Africans have embraced digital entertainment in recent years, most notably in Nigeria and South Africa, two of the continent's most sophisticated markets. Internet-based video content has truly taken off in Africa, competing with broadcast TV for consumers' attention. The growing popularity of digital entertainment among African consumers can be attributed to better network coverage and the wider availability of affordable smartphones. African consumers increasingly want to create their own viewing/listening mix when it comes to digital entertainment. Figure 7.5 shows a group of women enjoying digital entertainment [15].

– *Tourism:* Tourism is a social, cultural, and economic phenomenon that entails the movement of people to places outside their usual environment. Although information and communication technologies (ICT) have been adopted in some aspects of the tourism industry, there is further room to enhance the performance of this industry through the adoption of IoT technologies. For example, the mission of the South African Department of Tourism is to create a conducive environment for growing and developing tourism through innovation, strategic partnerships, the provision of information, and knowledge management services. South Africa is endowed with archeological sites, arts and culture sites, botanical gardens, caves, historical sites, museums, natural wonders, waterfalls, world heritage sites, blue flag beaches, etc., for tourism. IoT has a place in the monitoring and tracking of wild animals. The adoption of IoT for the South African tourism industry enhances the efficiency of the industry and impacts the South African economy [16].

Figure 7.5: A group of women enjoying digital entertainment [15].

7.4 Adapting Internet of Things in African Nations

In Africa, we are seeing an increased adoption of IoT technology, which is specifically being used to transform the lives of individuals on the continent. African countries such as Ghana, Nigeria, Rwanda, and South Africa have seen a steady rise in successful IoT implementations meant to improve key areas of sustainable development – water monitoring being one of the most popular sectors. Figure 7.6 displays an official checking an Internet-based water monitoring device at a borehole in Burkina Faso [17]. We examine how IoT technology is being used in some selected African nations [18, 19]:

– *South Africa:* This country is taking the lead in terms of adopting the IoT. It is one of the fastest-growing IoT markets on the continent. The IoT boom in South Africa presents an array of opportunities across various sectors. IoT technology promises smarter cities, streamlined logistics, enhanced healthcare, and improved energy management. One of the key examples of how IoT is used in South Africa is electronic tolling systems, which are an innovative technology that uses sensors and digital connec-

Figure 7.6: An official checking an Internet-based water monitoring device [17].

Figure 7.7: The toll system on the highway in South Africa [19].

tions. These systems connect everything from traffic status and the number of ve-hicles passing through toll gates to electricity grids and traffic controls. The E-toll sys-tem collects tolls electronically without human intervention, as there are no physical booths on the highway. This tolling system utilized IoT technology to make toll pay-ments more efficient and cost-effective via an IoT-based electronic tag. Figure 7.7 shows the toll system on the highway [19]. Headquartered in South Africa, Africa Wildlife Tracking (AWT) is using ORBCOMM's state-of-the-art satellite modems to pro-vide secure, near real-time GPS tracking and monitoring of large animals such as ele-phants in some of the world's most remote regions and densest forests. The ORB-COMM modem's small size and low-power consumption transceivers have resulted in improved longevity and performance in the battery-powered elephant collars [20, 21].

– *Nigeria:* This country is the biggest mobile market and the most populous African country. It has enormous prospects in IoT, which, if effectively implemented, is likely to bring about increased productivity across all economic sectors and an improved stan-dard of living for Nigerians. In Nigeria, IoT is being used for monitoring and controlling energy consumption in buildings and for tracking and managing medical equipment. The National Agency for Food and Drug Administration and Control, in 2010, resorted to the IoT-based product verification initiative to curb drug counterfeiting by using RFID. The IoT solution used tags equipped with RFID to secure the integrity of the drugs throughout the supply chain. Figure 7.8 shows how RFID tags are used to prevent coun-terfeit drugs [19]. One of the most important applications of IoT in Nigeria was the use of RFID cards and readers in the 2015 general elections. The technology was used to check the authenticity of voters in the elections and greatly improved the credibility of the process by its ability to detect fake and cloned permanent voter cards, thus curbing massive thumb printing. Another important application of IoT technology in Nigeria is the use of unmanned aerial vehicles in the fight against terrorism [22]. IoT Africa is the

Figure 7.8: RFID tags are used to prevent counterfeit drugs [19].

Sigfox operator for Nigeria. Sigfox in Nigeria will build the low-power wide-area network and allow connectivity of Sigfox-compliant devices to connect to the network. With IoT Africa as a Sigfox, the Nigerian economy should expect a substantial boost. IoT Africa is aiming to offer a wide range of high-quality connected products such as utility meters, gas sensors, health watches, smart IDs, smoke detectors, container trackers, and street lighting solutions [23].

– *Kenya:* In Kenya, IoT is being used for smart city initiatives, including traffic management and waste management. Nairobi County in Kenya has been struggling with waste management challenges for a long time. In Kenya, an IoT-based application is being used to manage waste in an efficient and cost-effective way. It is also meant to create a digital map of Nairobi's streets. Some companies have developed sensor-based systems that can detect the amount of waste present in bins across the country. The system has been proven to reduce costs by up to 40%, making it an efficient and environmentally friendly way of dealing with waste in Kenya.

– *Tanzania:* In Tanzania, IoT technology is being used to stop oil pilferage within fleets via RFID technology. An IoT-enabled application using RFID was implemented to track each truck within the fleet in real-time. The application involved attaching an IoT-enabled gateway device to the truck's cabin area and RFID-enabled tags to the hatch. Usangu Logistics is a heavy transport company with a fleet of over 100 trucks and tankers dedicated to serving thousands of customers in Tanzania with oil, lubricants, and other bulky products. One such truck is shown in Figure 7.9 [19]. An IoT-enabled gateway device is attached to the truck's cabin area, and the seals are tagged with RFID-enabled tags. The implementation of this IoT-enabled solution resulted in a very significant drop in cases of pilfering of the oil that the trucks and tankers were carrying.

Figure 7.9: A truck for transporting bulky products [19].

– *Egypt:* In Egypt, IoT technology is being used to control household appliances and improve the energy efficiency of homes. Integreight developed an IoT chip that can be integrated with modern appliances like refrigerators, cameras, TVs, and washing machines, allowing users to control their appliances remotely via their smartphones.

– *Namibia:* This small African nation is not to be left behind in the field of IoT. In order to improve the effectiveness of antiretroviral drugs, Namibia implemented an IoT-based EDT. Pharmacists use this tool to dispense the correct medicine in the correct amounts to patients. The EDT is presented in Figure 7.10 [19].

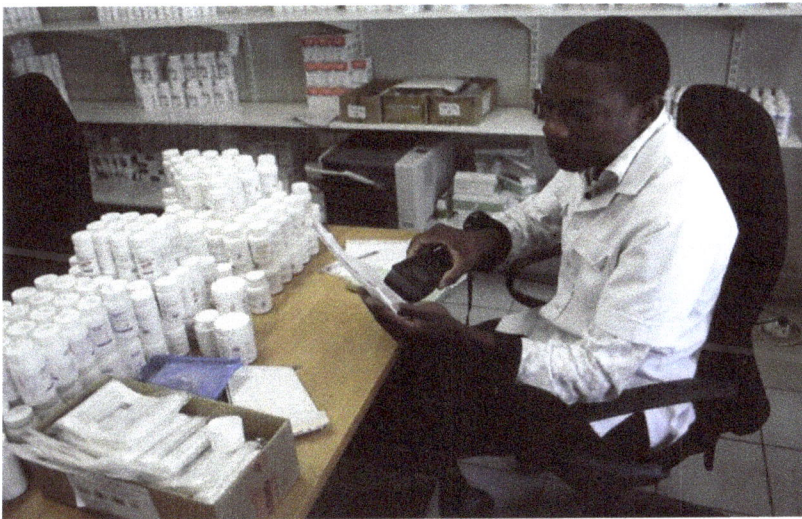

Figure 7.10: The electronic dispensing tool (EDT) [19].

7.5 Benefits

The IoT is a distributed network of smart devices that offers an ecosystem of business and application areas such as transportation, education, security, utilities, service, health, and quality delivery. The opportunities presented by IoT in Africa are huge. The benefits of IoT across a wide variety of sectors include improved access to essential resources, asset tracking, security solutions, enhanced healthcare, energy optimization, fleet management, waste management, and EV charge point infrastructure. The adoption of IoT technology is being used to transform the lives of individuals on the continent. Other benefits of IoT in Africa include the following [18, 24, 25]:

– *Connectivity:* In the digital world, connectivity defines convenience and efficiency. Due to the expansion of trade and transport networks, companies are working more internationally. Goods and services need to be monitored and secured while traveling vast distances. Devices require international connectivity, roaming, and built-in resilience.

– *Economic Growth:* A key benefit of IoT is that it is being used to help many African countries drive rapid economic growth. There are several areas that are seeing high growth due to IoT. The technology is being used to help many African countries drive rapid economic growth. Creating new jobs and helping businesses become more competitive in the global marketplace. As the IoT market continues to expand and evolve, Africa holds immense potential for growth and innovation. The IoT applications being adopted are focused on bettering the daily lives of those living on the continent and those aimed at developing economies.

– *Smart Agriculture:* Agriculture is the backbone of many African economies, yet farmers in remote areas face numerous challenges. IoT technology is revolutionizing agriculture in Africa. It is providing farmers with actionable insights. Sensors can monitor soil moisture, weather conditions, and crop growth, enabling farmers to make informed decisions. IoT is useful in assisting the development of precision agriculture in Africa. There are still several IT-related issues that hinder the advancement of IoT in agriculture. These include expensive equipment, the complexity of the systems, and the lack of established sensor network standards.

– *Healthcare:* Access to healthcare is a fundamental human right, yet millions of people in rural Africa lack basic medical facilities. IoT is being used for healthcare solutions that provide better diagnosis and remote treatment options for patients. IoT has the potential to enable remote patient monitoring and telemedicine services. IoT-based smart sensors can remotely monitor patients' vital signs and chronic conditions. RFID tags track medical equipment and supplies in the healthcare sector. Wearable sensors have been widely used in the healthcare industry. Body sensing networks are one of the areas that have been proposed for the efficient use of IoT in healthcare.

– *Smart Living:* IoT technology improves lives and allows for smart living. One can control and monitor smart IoT appliances remotely through smartphone apps or web interfaces. These smart IoT appliances offer several benefits that enhance convenience, efficiency, and connectivity in various aspects of an individual's daily life. They allow users to check the status, adjust settings, and receive notifications from anywhere, providing greater convenience and peace of mind.

– *Urban Management:* African cities are using IoT to become smarter and more efficient. From traffic management systems that reduce congestion to smart grids that optimize energy use, IoT is helping address urban challenges and improve the quality of life for people living in cities.

– *Retail:* African retail stores utilize RFID technology for inventory management, preventing theft, and improving the shopping experience. IoT technologies assist businesses and organizations with tracking and managing assets and vehicles and improving overall efficiency.

– *Clean Energy:* One of the most significant challenges in off-grid living is access to reliable energy sources. IoT-enabled solar systems are changing things. Smart solar panels, equipped with sensors, can take advantage of Africa's abundance of sunshine and optimize energy production and distribution. IoT devices can monitor energy consumption patterns, enabling users to manage their energy usage effectively.

– *Water Management:* Access to clean water is another pressing issue in most regions of Africa. IoT-powered water management systems offer a sustainable solution. Smart sensors installed in water pumps and distribution networks can monitor water levels, quality, and leaks in real-time. This data allows for proactive maintenance and efficient use of water resources.

– *Disaster Management:* Natural disasters, such as floods and droughts, are frequent occurrences in Africa. IoT technology plays a crucial role in disaster preparedness and response. Early warning systems equipped with sensors and predictive analytics can forecast impending disasters and alert communities at risk.

7.6 Challenges

The adoption of IoT in Africa has been slow due to challenges such as cost, accessibility, shortage of experienced talent, infrastructure deficiencies, high levels of poverty, underdeveloped economies, and poor insights. Other challenges include the following [26–31]:

– *Security and Privacy:* The two main obstacles to IoT are security and privacy. The extent of these issues can be very critical when we imagine the hacking of power grid stations and petroleum refineries. Privacy and security are particularly critical in wearable healthcare devices, and IoT needs to be highly protected against privacy breaches and malicious attacks that might cause harm to users. Security is one of the critical issues among the various IoT technologies within network communication such as 2G, 3G, 4G, optic networks, or WSNs. The IoT is facing security issues, increasing susceptibility to vulnerabilities, attacks, and threats to infrastructure, applications, and services via the Internet. The increase in security vulnerabilities, threats, and attacks has raised eyebrows in Africa in both private and public firms. The major issue experienced in big data and IoT is security, as a result of computing and storing large amounts of datasets within cloud storage.

– *Connectivity:* In order for IoT to work effectively, it relies on high-speed Internet connections. One of the main challenges in certain regions of Africa is the lack of adequate Internet connectivity. Some regions have limited mobile connectivity with few operators. Rural communities are unable to benefit from the advantages that IoT can offer. The challenge then becomes providing reliable mobile connectivity across diverse and sometimes remote areas. The availability of sufficient and economical infrastructure and Internet bandwidth for implementing IoT solutions is essential, and governments in Africa should be able to provide this infrastructure to support the creation of IoT solutions.

– *Power Supply:* The unreliability, unavailability, and unaffordability of connectivity and power grid infrastructure hinder the implementation of a multi-layered IoT architecture for rural societal services. Due to the fact that things move around and are not connected to a power supply, their smartness needs to be powered by a self-sufficient energy source. Most IoT devices have small batteries. Most batteries and power packs are either too heavy, thereby making the entire system bulky, or they have a short lifespan and require frequent replacement or charging. Solar energy is set to become the biggest trend. The energy challenge is not complete without central processing units (CPUs). CPU consumption got heightened by the rising number of IoT-enabled devices signaling and sending data between one another.

– *Borderless Continent:* One of the most significant challenges in Africa is creating a "borderless continent" in the sense of connectivity. The sheer size of the continent, coupled with a disparate technology ecosystem, creates challenges for connecting devices. Multiple mobile network operators (MNOs) exist in Africa, but not a single MNO covers an entire country. Working with multiple MNOs creates operational complexity.

– *Regulation:* Africa lacks a uniform data governance framework. A lack of comprehensive IoT-specific regulations affects data security and privacy. An African enterprise solution must factor in the variety of regulatory and legal requirements across the continent. Comprehensive IoT-specific regulations are needed to foster market confidence.

– *Interoperability:* Different industries today use different standards to support their applications. With numerous sources of data and heterogeneous devices, the use of standard interfaces between these diverse entities becomes important. If manufacturers are to realize the promise and potential of IoT, it is critical that the billions of things that make up IoT are able to connect and interoperate. It is with that goal in mind that organizations such as the Open Interconnect Consortium and the Industrial Internet Consortium have been established.

– *Skills Gap:* There is a lack of skilled personnel and a shortage of IoT-specific training programs. African universities should teach about the IoT so that we create a critical mass of people who can meaningfully contribute in the evolution of the sector.

– *Architecture:* IoT includes a range of complex, integrated smart technologies that are mobile, transparent, decentralized, and invisible. This creates complexity in IoT data integrations of large datasets through heterogeneous environments. These necessitate heterogeneous architectures and hardware that are flexible and open-standard in IoT.

– *Compatibility:* This is due to heterogeneous technologies required to connect to suitable devices. This necessitates additional software and hardware due to incompatibilities and complexity.

– *Digital Illiteracy:* There has been significant progress in increasing adult literacy rates across sub-Saharan Africa in recent years. Despite these advances, around 37% of the adult population still lacks basic literacy skills, equivalent to over 170 million people. In addition to basic literacy, digital literacy, the ability to effectively and critically navigate, evaluate, and create information using a range of digital technologies, is also significantly lacking.

– *Digital Divide:* The digital divide is an injustice that must be addressed in Africa. The digital divide is a notable challenge in Africa, with many people still without access to ICTs. The advantages of the digital revolution have not been evenly distributed, with several regions in Africa still facing major barriers to accessing and utilizing ICTs. The digital divide, characterized by disparities in access, affordability, and digital literacy, poses a major hurdle to inclusive development in Africa. Access to technology alone is not sufficient to bridge the digital divide; digital literacy and skills development are equally important.

To address these challenges, collaborative partnerships between government, industry stakeholders, and research institutions are crucial.

7.7 Conclusion

The IoT is a system of interconnected computing devices, machines, objects, animals, or people that are provided with the ability to transfer data over a network without requiring human interaction. The network of devices is capable of collecting data, sharing data over the Internet, and sourcing information for users with ease. IoT technology has the potential to offer new jobs and provide new solutions to issues such as water and power shortages. IoT is transforming the way we live and work [32]. Every area of life stands to benefit from the innovations and efficiencies possible in a fully connected world. If explored quickly, harnessed, and implemented, IoT could be the great breakthrough for African industry and could put Africa at the forefront of this revolution.

The IoT is a reality that cannot be ignored. Although Africa is still behind the rest of the world in terms of Internet penetration, the gap is quickly closing. In spite of many obstacles to overcome, the future of IoT in Africa is bright and promising, with significant potential to transform various sectors such as healthcare and agriculture. For more information about the IoT in Africa, one should consult the following related journals:

- *IEEE Internet of Things Journal,*
- *Internet of Things; Engineering Cyber Physical Human Systems*

References

[1]   W. Shafik, "Chapter 7 navigating emerging challenges in robotics and artificial intelligence in Africa," https://www.irma-international.org/viewtitle/339985/?isxn=9781668499627
[2]   "New technology trends for 2022," December 2021, Unknown Source.
[3]   S. Bassendowski, "Technology in education column," *Canadian Journal of Nursing Informatics*, vol. 13, no. 1, March 2018.
[4]   M. M. Asad et al., "Investigating the impact of IoT-Based smart laboratories on students' academic performance in higher education," *Universal Access in the Information Society*, 2022.
[5]   M. N. O. Sadiku, U. C. Chukwu, and J. O. Sadiku, "Internet of things in education: A brief overview," *International Journal on Integrated Education*, vol. 6, no. 5, May 2023, pp. 304–312.
[6]   "Tata Communications transformation services and smart Africa alliance come together to bridge skills gap in Africa," July 15, 2019, Unknown Source.
[7]   M. N. O. Sadiku, U. C. Chukwu, and J. O. Sadiku, "Internet of things in Africa," *Excellencia: International Multi-disciplinary Journal of Education*, vol. 2, no. 5, 2024, pp. 1054–1066.
[8]   A. Ghosh, "Internet of things: Basics," March 2014, https://thecustomizewindows.com/2014/03/internet-things-basics/
[9]   M. N. O. Sadiku, S. M. Musa, and S. R. Nelatury, "Internet of things: An introduction," *International Journal of Engineering Research and Advanced Technology*, vol. 2, no. 3, March 2016, pp. 39–43.
[10] A. M. Rahmani et al., "E-Learning development based on internet of things and blockchain technology during COVID-19 pandemic," *Mathematics*, vol. 9, 2021.
[11] S. Anand and S. K. Routray, "Issues and challenges in healthcare narrowband IoT," *International Conference on Inventive Communication and Computational Technologies*, 2017, pp. 486–489.
[12] "Internet of things – Africa," https://www.statista.com/outlook/tmo/internet-of-things/africa

[13] A. Oliveira-Jr et al., "IoT sensing platform as a driver for digital farming in rural Africa," *Sensors* (Basel), vol. 20, no. 12, June 2020.

[14] D. Hattingh, "Top 5 IoT innovations transforming lives in Africa," October 2023, https://telecoms. adaptit.tech/blog/top-5-iot-innovations-transforming-lives-in-africa/

[15] "Africa & the Internet," https://news.africa-business.com/post/africa–the-internet-

[16] O. Gcaba and N. Dlodlo, "The Internet of things for South African tourism," https://web.archive.org/ web/20160910064504id_/http://researchspace.csir.co.za:80/dspace/bitstream/10204/8674/1/Gcaba_ 2016.pdf

[17] I. Atanga, "The Internet of everything water," May-July 2017, https://www.un.org/africarenewal/mag azine/may-july-2017/internet-everything-water

[18] "IoT in Africa: 4 Key use cases," https://caburntelecom.com/iot-in-africa/

[19] "6 IoT applications that improved people's lives in Africa – A story of 6 countries," https://www.vizo com.com/ict/6-iot-applications-that-improved-peoples-lives-in-africa-a-story-of-6-countries/

[20] G. Jack, "The IoT boom in South Africa: Opportunities and challenges," December 2023, https://www.linkedin.com/pulse/iot-boom-south-africa-opportunities-challenges-jack-geldenhuys-kausc#:~:text=The%20IoT%20boom%20in%20South%20Africa%2C%20projected%20to%20reach% 20a,healthcare%2C%20and%20improved%20energy%20management.

[21] "ORBCOMM's IoT technology and Africa wildlife tracking track conservation efforts around the world," October 2020, https://news.satnews.com/2020/10/08/orbcomms-satellite-iot-technology-and-africa-wildlife-tracking-support-conservation-efforts-around-the-world/

[22] M. Ndubuaku and D. Okereafor, "State of Internet of things deployment in Africa and its future: The Nigerian scenario," *The African Journal of Information and Communication*, vol. 15, no. 15, December 2015.

[23] "IoT Africa, Sigfox seal partnership deal," https://www.iotafricanetworks.com/post/iot-africa-sigfox-seal-partnership-deal

[24] "Top 5 IoT innovations transforming lives in Africa," https://telecoms.adaptit.tech/blog/top-5-iot-innovations-transforming-lives-in-africa/

[25] "The role of IoT in empowering off-grid living in Africa," https://worldov.com/insights/the-role-of-iot-in-empowering-off-grid-living-in-africa#:~:text=Smart%20solar%20panels%20equipped%20with,man age%20their%20energy%20usage%20effectively.

[26] "Closing the digital divide in Africa: Unfolding challenges, strategies, and success stories," March 2024, https://www.thecable.ng/closing-the-digital-divide-in-africa-unfolding-challenges-strategies-and-success-stories/

[27] R. Cohen, "The business case for IoT in Africa: Opportunities and challenges," https://flolive.net/ blog/the-business-case-for-iot-in-africa-opportunities-and-challenges/

[28] E. Umeh, "Digital transformation in Africa requires homegrown solutions," December 2021, https://hbr.org/2021/12/digital-transformation-in-africa-requires-homegrown-solutions?utm_me dium=paidsearch&utm_source=google&utm_campaign=domcontent&utm_term=Non-Brand&tpcc=paidsearch.google.dsacontent&gad_source=1&gclid=EAIaIQobChMIo-2ymcWShgMVi4 JaBR3AOgyPEAAYASAAEgKrSvD_BwE

[29] J. K. Machh et al., "Towards a strategic application of IoT and big data for African societal solutions," *IST-Africa 2020 Conference Proceedings Miriam Cunningham and Paul Cunningham (Eds) IST-Africa Institute and IIMC*, 2020.

[30] S. A. Isma'ili et al., "African societal challenges transformation through IoT," https://ro.uow.edu.au/ cgi/viewcontent.cgi?referer=&httpsredir=1&article=1665&context=eispapers1

[31] M. Ndubuaku and D. Okereafor, "Internet of things for Africa: Challenges and Opportunities," *Proceedings of International Conference On Cyberspace Governance*, November 2015, pp. 23–31.

[32] J. Stewart, "Challenges surrounding IoT deployment in Africa," December 2019, https://www.compar ethecloud.net/articles/challenges-surrounding-iot-deployment-in-africa/

Chapter 8
Blockchain in Africa

I believe that Blockchain will do for trusted transactions what the Internet has done for information.
– Ginni Rometty

8.1 Introduction

Africa is a continent that has 54 countries, with an area of 30,370,000 km^2 and 1.4 billion individuals as of 2021, subdivided into 5 major regions, like Northern Africa (with countries like Libya, Egypt, North Sudan, Algeria, Morocco, and Tunisia as demonstrated) inhabiting the northerly region of Africa [1]. The continent is not just catching up with the world; it is propelling itself to the forefront of innovation. Africa is rising, and its tech scene is leading the way. Africa is closely watched as the next big growth market. It is home to some of the youngest populations in the world and also to many fast-growing economies. Africa is a booming continent with incredible growth potential, as the second-largest continent in the world and the world's largest free trade area, connecting 1.3 billion people (16.6% of the world population) across 55 nations. The art and culture of Africa are diverse, reflecting the varied ethnic groups that inhabit the continent. Today's Africa is bogged down by insecurity, healthcare, corruption, lack of mechanized farming, economic instability, underdevelopment, poverty, unemployment, nepotism, and lingering effects of destructive colonialism, etc.

Contracts, transactions, and their records are critical, defining structures in our economic, legal, and political systems, but they have not being able to keep up with the world's digital transformation. Blockchain (BC) promises to solve this problem. BC (also known as "distributed ledger technology") is a peer-to-peer network that sits on top of the Internet. It was introduced in 2008 as part of a proposal for Bitcoin. Bitcoin is the first application of BC technology. Bitcoin is a cryptographic electronic payment system that purports to be the world's first cryptocurrency. It has become the most talked-about cryptocurrency. Figure 8.1 displays coins of various cryptocurrencies [2]. Bitcoin is completely open source so that any developer can download it, modify it, and create his own version of the software. This unique feature has led to an explosion of alternative Bitcoin implementations, popularly known as altcoins. Some of the popular implementations include IxCoin, Namecoin, Litecoin, Ripple, Dogecoin, and Bitcoin. Some of the key benefits of Bitcoin include security, transparency, lower transaction costs, anonymity, and resilience. Although Bitcoin is a revolutionary idea, its implementation suffers from some problems such as instability, deflation, lack of replicability, computational inefficiency, and lack of regulation or enforcement [3].

https://doi.org/10.1515/9783112211984-008

Figure 8.1: Coins of various cryptocurrencies [2].

The BC could bring everything that is good about Bitcoin and translate it into decentralized applications. BC refers to new applications of a distributed database technology that build on tamper-proof records of time-stamped transactions. By decentralizing it, BC makes data transparent to everyone involved, and this eliminates the risks that come with data being held centrally. A BC facilitates secure online transactions [4].

In recent years, BC technologies have been finding practical applications across the African continent and accelerating Africa's transition to a single digital economy. BC technology continues to impact Africa to a degree that is unprecedented anywhere else in the world. The African BC industry is understandably experiencing rapid growth, with investors readily embracing it. The transformative potential of this technology for business operations in Africa is undeniable. Nigeria is the leading country in terms of the number of BC startups funded, followed by South Africa, Seychelles, and Kenya [5].

African economies have been held back from their full potential, but BC technology could help change that. BC technology has been gaining a lot of attention from businesses, investors, and governments worldwide. It is having an influence on African lives, well-being, and resilience. It has come to be perceived as a groundbreaking method that can bypass the several systemic flaws affecting the continent. The increase in BC adoption demonstrates the rising confidence in the potential of BC technology to drive financial independence, infrastructure development, personal identification, and record-keeping in Africa [6].

This chapter examines the use or adoption of BC technology in Africa and its potential benefits for the region. It begins with describing what BC is all about. It discusses the uses of BC in Africa. It covers African nations that have adapted BC. It highlights the benefits and challenges of BC in Africa. The last section concludes with comments.

8.2 What Is Blockchain?

BC, a type of distributed digital ledger technology (DLT), is a relatively new and exciting way of recording transactions in the digital age. It is a decentralized and distributed DLT that securely records and verifies transactions across multiple computers or nodes in a network. Basically, it is a chain of blocks in which each block contains a list of transactions. BC technology was created as the foundational basis for Bitcoin – a digital currency in which secure peer-to-peer transactions occur over the Internet. It is expected that spending on BC solutions worldwide would grow from 4.5 billion USD (2020) to an estimated value of 19 billion USD by 2024 [7].

Originally developed as the accounting method for the virtual currency Bitcoin, BCs are appearing in a variety of commercial applications today. BC technology is a type of distributed digital ledger that uses encryption to make entries permanent and tamper-proof and can be programmed to record financial transactions. It is used for the secure transfer of money, assets, and information via a computer network such as the Internet without requiring a third-party intermediary. It is now being adopted across financial and non-financial sectors. As a catalyst for change, BC technology is going to change the business world and financial matters in major ways.

The first BC was conceived in 2008 by an anonymous person or group known as Satoshi Nakamoto, who published a white paper introducing the concept of a peer-to-peer electronic cash system he called Bitcoin [8, 9]. Bitcoin and Ethereum are the first two mainstream BCs. Other modern BCs include Namecoin, Peercoin, Ether, and Litecoin. Figure 8.2 shows different components of BC [10], while Figure 8.3 displays different types of BC [11].

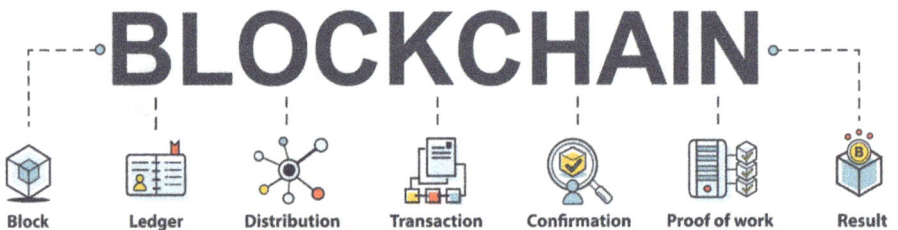

Figure 8.2: Different components of blockchain [10].

BC combines existing technologies such as distributed digital ledgers, encryption, immutable records management, asset tokenization, and decentralized governance to

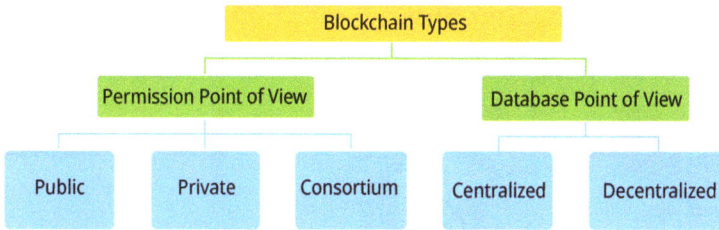

Figure 8.3: Types of blockchain [11].

capture and record information that participants in a network need to interact and transact. As illustrated in Figure 8.4, a complete BC incorporates all the following five elements [12]:

- *Distribution:* Digital assets are distributed, not copied or transferred. A protocol establishes a set of rules in the form of distributed mathematical computations that ensure the integrity of the data exchanged among a large number of computing devices without going through a trusted third party. A centralized architecture presents several issues, including a single point of failure and problems of scalability.
- *Encryption:* BC uses technologies such as public and private keys to record data securely and semi-anonymously. Completed transactions are cryptographically signed, time-stamped, and sequentially added to the ledger.
- *Immutability:* The BC was designed so these transactions are immutable, i.e., they cannot be deleted. No entity can modify the transaction records. Thus, BCs are secure and meddle-free by design. Data can be distributed, but not copied.
- *Tokenization:* Value is exchanged in the form of tokens, which can represent a wide variety of asset types, including monetary assets, units of data, or user identities.
- *Decentralization:* No single entity controls a majority of the nodes or dictates the rules. A consensus mechanism verifies and approves transactions, eliminating the need for a central intermediary to govern the network.

BC is a distributed ledger technology that evolved from the Internet of information and represents a second phase of the Internet. It is somewhat similar to spreadsheets or databases because it is a database where information is entered and stored. BC is a decentralized form of record-keeping. The key difference between a traditional database (or spreadsheet) and a BC is how the data is structured and accessed.

The term "blockchain" refers to the way BC stores transaction data – in "blocks" that are linked together to form a "chain." The chain grows as the number of transactions increases. A block is created whenever a transaction is made. Each transaction, referred to as a "block," is secured through cryptography, time-stamped, and val-

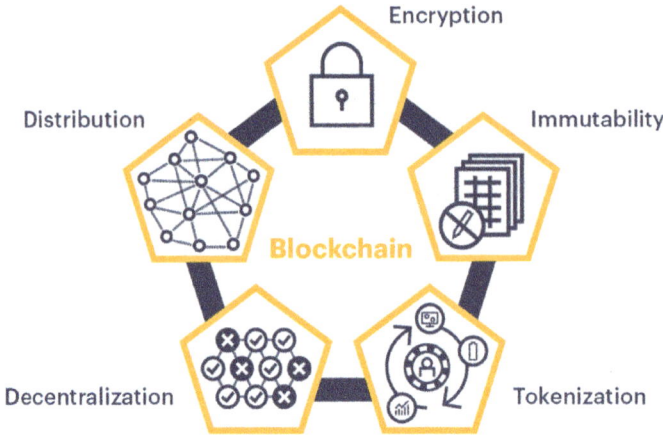

Figure 8.4: Five key elements of blockchain [12].

idated by every authorized member of the database using consensus algorithms. Every transaction is attached to the previous transaction in sequential order, creating a chain of transactions (or blocks), as shown in Figure 8.5 [13]. In other words, BC technology is the next evolution of the Internet. A block is the "current" part of a BC, which records some or all of the recent transactions. Each BC block has a unique 3s-bit whole number called a nonce, which is connected to a 256-bit hash number attached to it. The block is broadcasted to all nodes for validation. Once completed, a block goes into the BC as a permanent database. Each time a block gets completed, a new one is generated. Each data item in a BC has a timestamp. A BC is an ordered chain of blocks. All data of a transaction are traceable based on the chain structure of BC. Figure 8.6 displays how BC works [14].

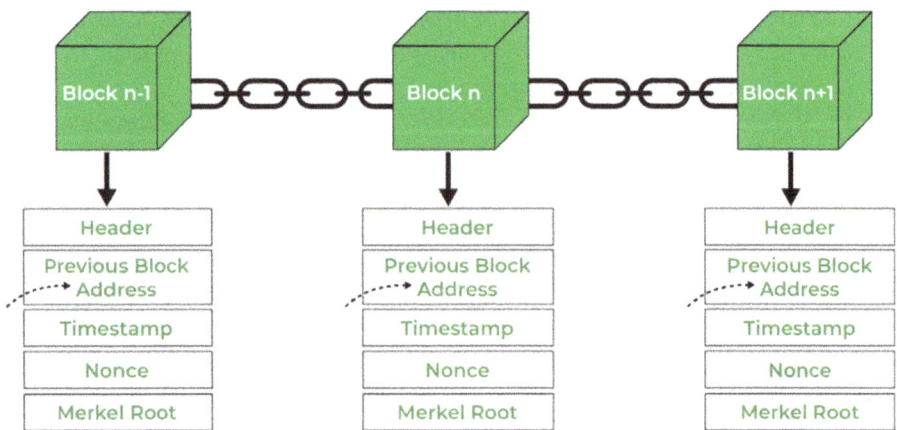

Figure 8.5: A chain of transactions (or blocks) [13].

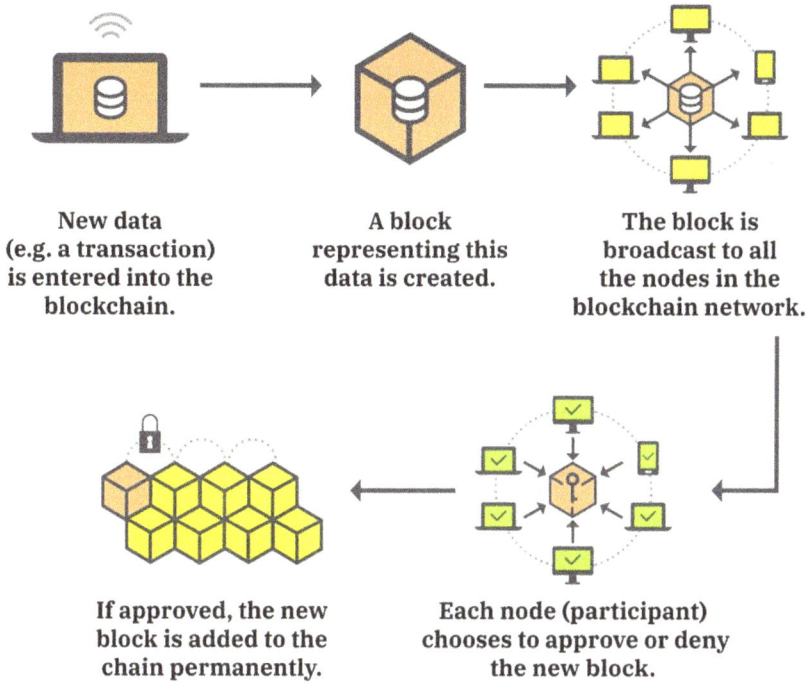

Figure 8.6: How blockchain works [14].

BC technology is the next evolution of the Internet. BC Africa has been a trending concept as a result of Africa's crypto adoption rate. BC Africa has significantly benefited from this mechanism since it works under the principle of not having a central location for data. BC Africa is illustrated in Figure 8.7 [15]. Nigeria, the current leading country in cryptocurrency, has also set its stage in BC Africa.

Figure 8.7: Blockchain in Africa [15].

8.3 Use of Blockchain in Africa

The lack of modern agriculture, underdevelopment, unemployment, and poverty has slowed Africa's development in the twenty-first century. BC technology in Africa offers alternatives to tackle its day-to-day problems. Some African countries have adopted BC technology, especially in the finance, energy, government, and agriculture sectors.

– *Finance:* BC's utility for financial applications has been well-established, and African BC companies often focus on payment systems and remittances. The main benefit of cryptocurrencies is not relying on third-party financial institutions. The African continent is on the brink of a financial transformation powered by BC technology. Africa, a continent known for its resourcefulness and adaptability, is undergoing a fintech revolution. Inflation and corruption have fomented mistrust in central banks and fiat currency; so many African consumers see BC-based cryptocurrencies as a promising alternative. With over 1.3 billion people spread across 55 countries, diverse challenges and opportunities abound. The traditional banking systems have attempted, without much success, to reach remote areas and serve unbanked populations. This has often led to poor economic growth. BC technology can potentially bridge the gap between traditional financial services and the unbanked in Africa [16]. Established financial institutions often come with numerous lending criteria and offer modest returns to lenders. However, African DeFi platforms like Finna Protocol are transforming the methods of lending and borrowing money. Financial institutions in Africa are becoming aware of BC technology's role in improving financial services such as credit, loans, and payments. Digital payments have become a very important part of modern African society. A major problem financial institutions face in Africa is a lack of infrastructure.

– *Energy:* In Africa, the demand for electricity largely exceeds supply. For example, Nigeria's shortage of 173,000 MW gave rise to large-scale imports of noisy and polluting power-generating sets. The cost of generation and distribution of power is high. Economic and political reasons mean it will take a lifetime for these communities to be considered. The UN Sustainable Development Goal #7, targeting universal energy access for all by 2030, is driving a global consensus on renewable energy in off-grid communities. With BC technology (that eliminates intermediaries), it is possible to establish an auditable, encrypted ledger that can record energy consumption, credit histories, as well as provide energy trading between households, giving consumers more control over their energy requirements and consumption. BC will inspire the fast adoption of decentralized energy systems in places with or without electricity. It will not only increase productivity among small energy consumers, but new ways of defining energy end-use will also emerge [17].

– *Government:* Some African governments attempt different measures to eliminate corruption, bad governance, mismanagement of public funds, and lack of accountability in their countries, but these efforts are mostly unsuccessful. BC can help in solving some of these problems, with accountability and integrity built in the technology by design. BC has been embraced at the level of civil society and is being addressed in political debates. BC technology has the potential to transform governance in Africa, particularly in the areas of transparency, efficiency, and trust, as well as improving governance and promoting economic growth. It is used in governance, including land registries, voting systems, and public finance management. Startups can play a pivotal role in putting pressure on corporations and governments and forcing them to dismantle obsolete frameworks [18]. A government-wide implementation of BC technology could potentially help drive a general behavioral change in society and drastically curtail systemic corruption and lack of accountability that plague Africa. Countries like Kenya and Nigeria are exploring BC for secure identity verification and reducing fraud in government programs.

– *Property Administration:* One of Africa's pressing issues has been land ownership disputes and fraudulent transactions. Real estate and land property ownership are usually not effectively managed by African governments, making land disputes very difficult to deal with. Due to corruption and nepotism, attempting to fix this issue has been largely unsuccessful in the past. Having a system that ensures each record of ownership is not only immutable solves the corruption problem [19]. The BC transformation makes keeping and securing records of real estate ownership, titles, products, and private equity shares simple. Land titles can become immutable by underpinning it with BC technology. That way, nobody can hack it.

– *Tax Administration:* With corruption being the primary reason and cause of under-collected tax returns in most African nations, BC technology could greatly benefit the economies, allowing the countries to collect billions in tax revenue that are being lost today. Because BC technology provides provenance, traceability, transparent, and immutable information about transactions, fraud and corruption are almost impossible in the system and easier to detect [20].

– *Healthcare:* For example, Kenya also intends to transform the health industry by introducing a BC-powered smart platform. Almost all public hospitals will now share a hub that facilitates data management, such as public resources and healthcare administrators.

– *Agriculture:* This is the backbone of many African economies. BC can revolutionize the sector by providing transparency and traceability in supply chains. African nations, with their rich natural resources and agricultural products, can use BC to ensure the authenticity and quality of products, reduce fraud, and improve trade efficiency. Agrikore, a Nigerian platform, uses BC to connect farmers, buyers, and suppliers, ensuring fair pricing and increasing food security across the continent.

8.4 Adapting Blockchain in African Nations

At the moment, Africa is one of the fastest-growing cryptocurrency markets in the world. Africa is a strong contender for developing technologies such as BC and crypto-currency due to the continent's growing mobile tech adoption rates. Some African countries (like South Africa, Nigeria, Zimbabwe, Kenya, and Ghana) have adopted the decentralized approach, while some countries (like Cameroon, Ethiopia, Lesotho, Sierra Leone, Tanzania, and the Republic of Congo) have banned crypto. The current situation of cryptocurrency in Africa is shown in Figure 8.8 [21]. We consider the following selected countries in Africa [22–25]:

– *South Africa:* South Africa has been one of the region's leaders in terms of crypto regulation and the development of supportive trading frameworks. The nation's pro-active approach to regulation has removed a lot of regulatory uncertainty. The predominant use case for crypto in South Africa revolves around investment. Citizens of the country have traded billions of dollars' worth of digital currency in recent years. In South Africa, the government is collaborating with others from the BRICS group of countries (which also includes Brazil, Russia, India, and China) on research into BC's potential for trade and other enablers of growth. Meanwhile, Standard Bank (the continent's largest financial institution) has joined Marco Polo, a BC-based trade finance network, which will help unlock access to BC finance throughout the 20 African countries where it operates.

– *Nigeria:* Nigeria boasts the largest population and economy in sub-Saharan Africa, as well as the largest cryptocurrency economy. BC has been one solution to Nigeria's economic challenges. Since 2016, Nigeria has suffered from an unstable political situation, the COVID-19 pandemic, and the collapse of oil prices. Thus, Nigerians are facing a high unemployment problem, causing some to migrate to other nations. Nigeria's uncertain economic environment has encouraged many citizens to seek financial alternatives, increasing the value proposition of cryptocurrency. In 2022, the nation co-operated with Bitt Inc. for its digital currency, eNaira, as shown in Figure 8.9 [25]. The move comes from the rising enthusiasm among companies and regulators throughout the continent to use distributed-ledger technology. Nigeria has also embraced BC technology in the education sector. The Nigeria Customs Service is currently looking into the possibilities of BC to enhance its operations. Nigeria has one of the most dynamic peer-to-peer Bitcoin trading markets in the world, but so far, the country has banned cryptocurrency.

– *Kenya:* This is East Africa's largest economy and is among the nations that have adopted BC technology in Africa. Kenya is already a hotbed of innovation in Africa. It has also been at the forefront of service digitization in the region. Kenya has already benefited from the adoption of M-PESA, the mobile money services. BitPesa, a money remittance network that converts digital currencies to local African currencies without involving third parties, has gained traction in Kenya. Kenya also intends to change the health industry by introducing a BC-powered smart platform and establish a fully automated Universal Patient Portal in Kenya. Consequently, AfyaRekod will provide patients and health personnel with real-time access to medical data and history through a secure central platform. In addition, Kenya's National Transport Safety Authority has also incorporated BC technology into its operations to revolutionize the transport industry. This will ultimately enable all vehicles to have electronic stickers on their windscreens, hence facilitating the recovery of stolen vehicles.

– *Ghana:* The Ghanaian government is set to become the first African country to use BC technology in its e-government processes. Since 2020, the Ghanaian government has raked 201 billion cedis in revenue from the services portal, money that could not be obtained initially due to bureaucratic hurdles and time-consuming processes. Ghana intends to introduce a digital currency known as e-Cedi, which will be the ultimate weapon in the fight against corruption. The central bank recently launched a regulatory "sandbox" that will allow banks, companies, and others to develop and pilot new BC-based products for merchant payments and remittances. In Ghana, Bitland has used BC to create a secure land registry, reducing land disputes and enhancing property rights. This innovation ensures that individuals have rightful land ownership, fostering economic growth and stability. The University of Ghana announced plans for a BC-based certificate management system.

– *Central African Republic:* This is the first country in Africa, and the second in the world after El Salvador, to designate Bitcoin as legal tender. The Bank of Central African States (BEAC) has banned the use of crypto for financial transactions in the Economic and Monetary Community of Central Africa (CEMAC), of which the Central African Republic is a member.

8.5 Benefits

Among the evangelists, there is a consensus that BC is an opportunity to catch up with developed economies by embracing a technology that would allow them to leapfrog and thwart the creation of soon-to-be-obsolete bodies. In spite of its seeming complexity, BC has clear real-world relevance for improved trade facilitation, especially in Africa. BC technology has the potential to help Africa overcome several developmental obstacles, such as poverty, a lack of financial inclusion, and a lack of trust in institutions. Efforts to harness BC for tangible, real-world benefits are increasing. Bitcoin, as the foremost and widely recognized cryptocurrency, is increasingly being adopted in Africa as a means of payment. BC's list of potential applications is endless. Other benefits of BC in Africa include the following [26–28]:

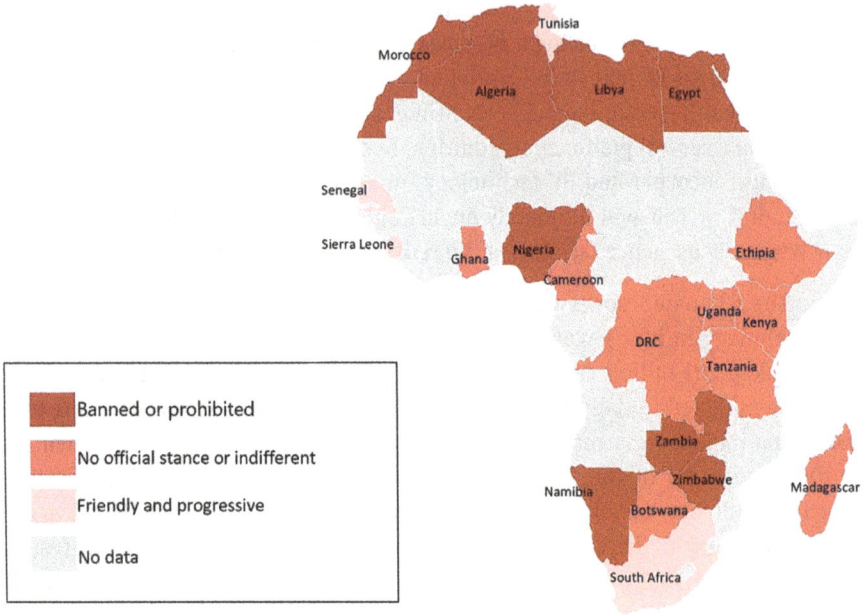

Figure 8.8: The current situation of cryptocurrency in Africa [21].

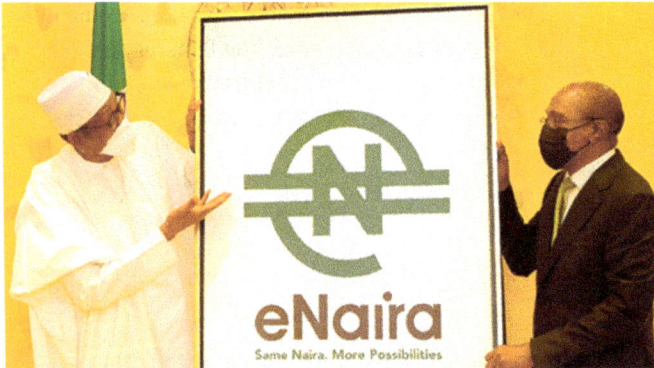

Figure 8.9: Electronic currency, eNaira, of Nigeria [25].

– *Privacy:* Data decentralization in distributed databases is very secure, as it comes with the concept of cryptography. BC helps in enhancing online privacies by allowing users to store their own digital footprints on their individual, unique BC, and control who can actually access them.

– *Financial Inclusion:* Through BC-based solutions, individuals can access financial services such as banking, lending, and remittances without the need for traditional banking infrastructure. DLT can facilitate financial inclusion by providing access to banking services for unbanked and underbanked populations.

– *Transparent Governance:* BC technology can enhance transparency and accountability in governance by creating immutable records of transactions and activities. Governments can use BC for voter registration, transparent procurement processes, and secure land registries, thereby reducing corruption and improving trust in public institutions.

– *Supply Chain Management:* DLT enables transparent and efficient supply chain management by tracking the movement of goods from production to consumption.

– *Cross-Border Payments:* BC facilitates cross-border transactions. Complex and unpredictable trade policies, as well as costly and time-consuming border procedures, continue to hold back African economies from reaching their full trading potential. BC technology can streamline cross-border payments and remittances by eliminating intermediaries and reducing transaction costs. African countries with high volumes of remittance inflows can leverage BC-based solutions to improve the efficiency and affordability of remittance services.

– *Identity Management:* BC-based digital identity solutions can help address identity fraud and provide individuals with secure identities. This is particularly important in Africa, where many people lack formal identification documents, hindering their access to essential services.

– *Tokenization of Assets:* Tokenization transforms physical assets into digital assets on the BC. BC enables the tokenization of assets, allowing individuals to fractionalize and trade assets such as real estate, commodities, and artwork. African countries can tokenize their assets to unlock liquidity, attract investment, and democratize access to investment opportunities.

– *Development Aid and Philanthropy:* BC technology can enhance transparency and traceability in development aid and philanthropy by providing donors with visibility into the use of funds and ensuring that aid reaches its intended recipients. This can help improve the effectiveness of aid programs and increase donor confidence.

– *Customs Clearance:* Implementation of the African trade agreement is still in its early stages, and customs processes vary across Africa. This red tape, typical of traditional customs systems, is a key impediment for small and medium-sized enterprises (SMEs) that are ready to enter global trading networks but do not yet have the financial resources to navigate the intricacies of bureaucracy. With BC, this complex process can be digitalized and streamlined, saving significant amounts of time and money.

– *Traceability:* The ability to accurately track cross-border shipments is crucial to the verifiability of standards and certifications, as well as the reliability and timeliness of delivery. Traditional methods of tracing are opaque, lack standardization, and are susceptible to interference and fraud. BC has shown great promise in increasing transparency to overcome these obstacles. Its open-source yet secure nature allows companies to assign and verify certifications easily. Already, BC is enabling a range of African products to reach end consumers through ethically verified supply chains.

– *E-Payments and Trade Finance:* In any given trade transaction, exporters, importers, and banks engage in a complex flurry of financial assurances and legal paperwork to show both sides that the other will make good on the deal. This paper-based system delays both the movement of goods and the payment cycle for such transactions. BC technologies streamline the ways organizations can track and verify the authenticity of such documentation, reducing transaction time and cost.

– *Low Costs of Remittance:* Under normal circumstances, when people wish to borrow from each other, they go through intermediaries like banks. With the introduction of BC, the use of third parties has been removed, leading to faster transaction processing.

– *Creation of Jobs:* The development of Bitcoin as a common currency in Africa has opened many doors for job creation in many companies and industries.

8.6 Challenges

All technologies have drawbacks, and BC is no exception. While the potential of BC in Africa is great, challenges persist. Regulatory uncertainty, lack of infrastructure, and limited awareness are some of the challenges that need to be addressed.
    Other challenges of BC in Africa include the following [5, 29–31]:

– *Scalability:* BC networks can be slow and inefficient due to the high computational requirements needed to validate transactions. As the number of users, transactions, and applications increases, the ability of BC networks to process and validate them in a timely manner becomes strained. Ensuring the ability of the BC network to handle a massive volume of transactions both rapidly and precisely is essential if it is to help address the scale of corruption in public procurement across Africa.

– *Energy Consumption:* The process of validating transactions on a BC network requires a lot of computing power, which, in turn, requires a lot of energy. This has led to concerns about the environmental impact of BC technology.

– *Lack of Standardization:* The lack of standardization arises issues such as interoperability, increased costs, and difficult mechanisms, making mass adoption an impossible task. This is acting as a barrier for the entry of new developers and investors as well.

– *Governance:* BC governance is still in its infancy, and there is no clear consensus on how to govern these decentralized systems. This lack of governance can lead to conflicts between stakeholders and make it difficult to implement changes or upgrades.

In markets with no regulatory bans, we tend to see the industry develop more responsibly as the market operates above ground, with more productive interaction between regulators and exchanges.

– *Security:* While BC technology is considered secure due to its decentralized nature, it is not immune to attacks. There have been several high-profile hacks and security breaches in recent years.

– *Privacy:* Privacy and the protection of sensitive information remain imperative. BC systems need robust safeguards against unauthorized access.

– *Infrastructure:* A major challenge is that several people in Africa do not have access to the Internet, and so they cannot use BC technology. For BC implementation, nations require foundational tech infrastructure, like a stable electricity supply, secure Internet connections, data storage and processing capacities, etc. Such infrastructure is scarce in many African countries at the moment. The continent is well known for its unstable power supply and low internet connectivity. BC can still be utilized in Africa even if the infrastructure needed to support it is not situated on the continent. All you need is a data-enabled mobile phone.

– *Knowledge Gap:* Introduction of BC would demand specialized expertise, which may currently be lacking. In spite of the rapid growth and progress made, there remains a significant knowledge gap on BC among corporate leaders and decision-makers. More corporate education on the benefits of BC technology is needed. This lack of knowledge and understanding hinders the potential for further growth and progress. It is imperative that we bridge this gap and educate those in positions of power. Efforts should be made to educate businesses, government officials, and the general public about the benefits of BC technology, including increased transparency, security, and efficiency.

– *Corruption:* Corruption and fraud are classic problems in Africa. Government expenditure on goods, services, and the funding of projects is plagued by corruption in Africa. The substantial amounts of money involved, the intricate processes, and sometimes questionable ties between officials and businesses make public procurement a fertile ground for corrupt practices. BC technology has emerged as a promising solution to reform public procurement and drastically curb corruption. BC's inherent transparency enables real-time oversight of procurement, ensuring each step is meticulously recorded, thereby eradicating opportunities for covert malpractices.

– *Regulation:* Regulating a highly volatile and decentralized system remains a challenge for most governments, requiring a balance between minimizing risk and maximizing innovation. Resistance from regulators is one of the biggest challenges of cryptocurrency in Africa. Any approach to regulate BC technology should commence with a clear consensus on regulatory objectives that are based on the particular positions of the governments involved.

– *Fear:* Some countries have banned cryptocurrency. Policymakers are worried that cryptocurrencies can be used to transfer funds illegally out of the region, and to circumvent local rules to prevent capital outflows. If crypto assets are accepted by the government as a means of payment, it could put public finances at risk.

– *Talent Shortage:* There is a massive shortage of the talent and resources required to build the solutions the continent needs to accelerate development. The current workforce needs to broaden its skillset. Since we have few BC experts on the continent, the logical thing to do is train more, but it is not easy.

– *Lack of Government Support:* Another notable problem is the lack of government support. We await a government-backed framework for BC technology.

In spite of these challenges, several African nations have already shown an inclination toward using BC. Initiatives in South Africa, Nigeria, and Rwanda illustrate the potentially transformative power of the new technology.

8.7 Conclusion

A BC is an incorruptible digital ledger. It has been heralded as a "game changer" for the development of African economies. BC technology offers distinct advantages over database technology as it provides for trustless recording of transaction data without relying on an existing intermediary. Although BC was originally used to create Bitcoin, today, it has universal applicability. As the technology continues to make inroads in global supply chains, governments, companies, and organizations have the chance to accelerate its adoption and reap the benefits of lower transaction costs, efficient delivery, increased exports, and more inclusive growth. With continued investment, collaboration, and education, Africa has the opportunity to become a global leader in BC technology. This also presents the potential for greater socioeconomic inclusiveness.

Africa has consistently been one of the smallest cryptocurrency markets or the smallest crypto economy among all regions. It is currently tackling challenges to technologies and innovation in the journey to embracing BC technology. BC has the potential to transform economic activity and improve living standards in Africa. BC technology in Africa remains crucial as the continent attempts to ascend the growth ladder. The future of cryptocurrency in sub-Saharan Africa looks bright and promising. For more information about BC in Africa, one should consult the following related journal: *Blockchain: Research and Applications.*

References

[1]   W. Shafik, "Chapter 7 navigating emerging challenges in robotics and artificial intelligence in Africa," https://www.irma-international.org/viewtitle/339985/?isxn=9781668499627

[2]   "U.S. companies and their cryptocurrency holdings," May 2022, https://www.reuters.com/business/finance/us-companies-their-cryptocurrency-holdings-2022-05-12/

[3]   A. Guadamuz and C. Marsden, "Blockchains and Bitcoin: Regulatory responses to cryptocurrencies," *Peer-reviewed Journal on the Internet*, vol. 20, no. 12, Dec. 2015.

[4]   M. N. O. Sadiku, Y. Wang, S. Cui, and S. M. Musa, "A primer on blockchain," *International Journal of Advances in Scientific Research and Engineering*, vol. 4, no. 2, February 2018, pp. 40–44.

[5]   "Blockchain in Africa: Prospects and problems over the next years," https://techgist.org/blockchain-in-africa-prospects-and-problems/

[6]   M. N. O. Sadiku, U. C. Chukwu, and J. O. Sadiku, "Blockchain in Africa," *Synergy: Cross-disciplinary Journal of Digital Investigation*, vol. 2, no. 5, 2024, pp. 96–110.

[7]   C. M. M. Kotteti and M. N. O. Sadiku, "Blockchain technology," *International Journal of Trend in Research and Development*, vol. 10, no. 3, May-June 2023, pp. 274–276.

[8]   "Blockchain," *Wikipedia*, the free encyclopedia https://en.wikipedia.org/wiki/Blockchain

[9]   S. Nakamoto, "Bitcoin: a peer-to-peer electronic cash system," https://bitcoin.org/bitcoin.pdf

[10] "The beginning of a new era in technology: Blockchain traceability," https://www.visiott.com/blog/blockchain-traceability/#:~:text=The%20Beginning%20of%20a%20New,money%20without%20a%20central%20bank.

[11] S. P. Mohanty et al., "PUFchain: Hardware-assisted blockchain for sustainable simultaneous device and data security in the Internet of everything (IoE)," *IEEE Consumer Electronics Magazine*, vol. 9, no. 2, March 2020, pp. 8–16.

[12] "The CIO's guide to Blockchain," https://www.gartner.com/smarterwithgartner/the-cios-guide-to-blockchain#:~:text=True%20blockchain%20has%20five%20elements,%2C%20immutability%2C%20tokenization%20and%20decentralization.

[13] "Blockchain structure," https://www.geeksforgeeks.org/blockchain-structure/

[14] G. O. R. Cruz, "What is blockchain?" June 2022, https://money.com/what-is-blockchain/

[15] "Blockchain Africa – the dawn of a new tech age," October 2022, https://furtherafrica.com/2022/10/20/blockchain-africa-the-dawn-of-a-new-tech-age/

[16] "Unlocking the potential of blockchain in Africa," September 2023, https://www.linkedin.com/pulse/unlocking-potential-blockchain-africa-fintech-association-of-kenya

[17] "Blockchain can revolutionise the energy industry in Africa," November 2018, https://www.weforum.org/agenda/2018/11/blockchain-will-change-the-face-of-renewable-energy-in-africa-here-s-how/#:~:text=With%20the%20emergence%20of%20blockchain,energy%20trading%20between%20households%3B%20giving

[18] "Powering the African vision – blockchain technology for Africa's transformative governance," August 2021, https://africanlii.org/akn/aa-au/doc/report/2021-08-31/powering-the-african-vision-blockchain-technology-for-africas-transformative-governance/eng@2021-08-31

[19] T. Samme-Nlar, "Blockchain technology can solve some of Africa's problems," September, 2020, https://gefona.org/blockchain-technology-can-solve-some-of-africas-problems/#:~:text=The%20accuracy%2C%20provenance%2C%20transparency%20and,tax%20administration%20could%20benefit%20greatly.

[20] "Cryptocurrency adoption in Africa," https://www.do4africa.org/en/cryptocurrency-adoption-in-africa/

[21] "Cryptocurrency penetrates key markets in sub-Saharan Africa as an inflation mitigation and trading vehicle," September 2023, https://www.chainalysis.com/blog/africa-cryptocurrency-adoption/

[22] "The increasingly significant role of blockchain technology in Africa," April 2023, https://furtherafrica.com/2023/04/02/the-increasingly-significant-role-of-blockchain-technology-in-africa/

[23] H. Fuje, S. Quayyum, and T. Molosiwa, "Africa's growing crypto market needs better regulations," November 2022, https://www.imf.org/en/Blogs/Articles/2022/11/22/africas-growing-crypto-market-needs-better-regulations#:~:text=The%20collapse%20of%20the%20world's,regulation%20of%20the%20crypto%20industry.

[24] B. Nzomo, "Ghana bets on blockchain to fight graft," May 2024, https://kenyanwallstreet.com/ghana-bets-on-blockchain-to-fight-graft/#:~:text=The%20Ghanaian%20government%20is%20set,country's%20vice%20president%20Mahamadu%20Bawumia.

[25] "Adoption of cryptocurrencies: Africa is gaining space," October 2023, https://resilient.digital-africa.co/en/blog/2023/10/17/adoption-of-cryptocurrencies-africa-is-gaining-space/

[26] "Africa blockchain," https://www.linkedin.com/pulse/africa-blockchain-belobaba-uyqre

[27] R. Benjelloun, "Opinion: How blockchain can boost trade in Africa," September 2021, https://www.devex.com/news/opinion-how-blockchain-can-boost-trade-in-africa-101615

[28] J. Jepkoech and C. A. Shibwabo, "Implementation of blockchain technology in Africa," *European Journal of Computer Science and Information Technology*, vol. 7, no. 4, August 2019, pp. 1–4.

[29] "African blockchain funding soars by 429%," https://african.business/2023/04/technology-information/african-blockchain-funding-soars-by-429#:~:text=African%20blockchain%20startups%20raised%20%24474,in%20association%20with%20Standard%20Bank.

[30] G. Ashiru, "Revolutionizing Africa with blockchain: Key applications and innovations," https://www.techinafrica.com/revolutionizing-africa-with-blockchain-key-applications-and-innovations/#google_vignette

[31] "Opportunities and challenges for the next decade," https://smartafrica.org/knowledge/data-scientist/

Chapter 9
3D Printing in Africa

3D Printing has the potential to revolutionize the way we make almost everything. – Barack Obama

9.1 Introduction

Africa is not just catching up with the world; it is propelling itself to the forefront of innovation. Africa is rising, and its tech scene is leading the way. Africa is closely watched as the next big growth market. It is the home to some of the youngest populations in the world and also to many fast-growing economies. Africa is a booming continent with incredible growth potential, as the second-largest continent in the world and the world's largest free trade area, connecting 1.3 billion people (16.6% of the world population) across 55 nations. Africa, a continent rich in resources and diverse cultures, has the potential to harness three-dimensional (3D) printing (3DP) technology to uplift local industries.

Traditionally, a printer is used at home or in the office to print out text and images on paper. This conventional printer prints in a flat two-dimensional (2D) space using the length and width dimensions. A 3D printer uses length and width but also adds depth to the print. A 3D printer has more manufacturing capacity than a traditional manufacturing machine. It has established itself not only as a disruptive technology but as a viable alternative to traditional manufacturing processes.

3DP is a type of industrial robotics. It uses raw material combination and builds an object one layer at a time. Advances in new material advances are keeping 3DP at the forefront of innovation across many industries such as healthcare, aerospace, engineering, printing, manufacturing, entertainment, education, chemistry, mathematics, biology, history, and architecture.

3DP is a technology that allows users to turn any digital file into a 3D physical object. It is the process for making a physical object from a 3D computer-aided design (CAD) file via a layering approach. It has been applied in diverse sectors, including architecture, construction, healthcare, manufacturing, engineering, clothing, jewelry, defense, and education. 3DP technology has brought a new wave of optimism and enthusiasm about the prospect of Africa's economic miracle. Some believe that 3DP presents significant opportunities for technological leapfrogging in African nations. As the world undergoes a digital revolution, Africa has the opportunity to leverage 3DP technology for sustainable development. 3DP is revolutionizing the way we manufacture products and is proving to be a game changer for Africa [1].

This chapter explores the state of art of 3DP technology in Africa. It begins with describing the concept of 3DP. It covers different types of 3DP. It discusses the uses of

https://doi.org/10.1515/9783112211984-009

3DP in Africa. It covers African nations that have adapted 3DP. It highlights the benefits and challenges of 3DP in Africa. The last section concludes with comments.

9.2 What Is 3D Printing?

3DP (also known as additive manufacturing (AM) or rapid prototyping (RP)) was invented in the early 1980s by Charles Hull, who is regarded as the father of 3DP. Since then it has been used in manufacturing, automotive, electronics, aviation, aerospace, aeronautics, engineering, architecture, pharmaceutics, consumer products, education, entertainment, medicine, space missions, the military, chemical industry, maritime industry, printing industry, and jewelry industry [2].

A 3D printer works by "printing" objects. Instead of using ink, it uses more substantive materials – plastics, metal, rubber, and the like. It scans an object – or takes an existing scan of an object – and slices it into layers, which can then convert into a physical object. Layer by layer, the 3D printer can replicate images created in CAD programs. In other words, 3DP instructs a computer to apply layer upon layer of a specific material (such as plastic or metal) until the final product is built. This is distinct from conventional manufacturing methods, which often rely on removal (by cutting, drilling, chopping, grinding, forging, etc.) instead of addition. Models can be multicolored to highlight important features, such as tumors, cavities, and vascular tracks. 3DP technology can build a 3D object in almost any shape imaginable as defined in a CAD file. It is additive technology as distinct from traditional manufacturing techniques, which are subtractive processes in which the material is removed by cutting or drilling [3].

3DP has started breaking through into the mainstream in recent years, with some models becoming affordable enough for home use. Many industries and professions around the world now use 3DP. It plays a key role in making companies more competitive. The gap between industry and graduating students can be bridged by including the same cutting-edge tools, such as 3DP, professionals use every day into the curriculum. There are 3D-printed homes, prosthetics, surgical devices, drones, hearing aids, and electric engine components. As shown in Figure 9.1, 3DP involves three steps [4]. A typical 3D printer is shown in Figure 9.2 [5].

9.3 Types of 3D Printing

There are generally three types of AM: selective binding, selective solidification, and selective deposition. Typically, people refer to these technologies as selective laser sintering (SLS), stereolithography (SLA), and fused deposition modeling (FDM), which are discussed as follows [6, 7]:

STEP 3
PRINTING

STEP 1
SCANNING

STEP 2
MODELLING

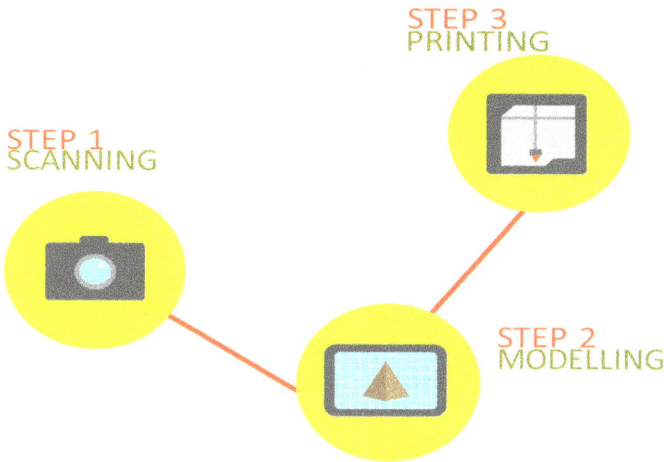

Figure 9.1: 3D printing involves three steps [4].

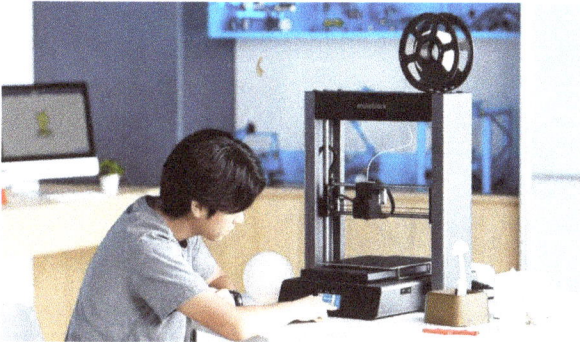

Figure 9.2: A typical 3D printer [5].

– *SLA*: This was the world's first 3DP technology, invented in the 1980s. It is an AM process that employs a vat of liquid ultraviolet curable photopolymer "resin" and an ultraviolet laser to build parts of layers one at a time. For each layer, the laser beam traces a cross section of the part pattern on the surface of the liquid resin. SLA parts have the highest resolution and accuracy and the smoothest surface finish of all plastic 3DP technologies. Although SLA can produce a wide variety of shapes, it has often been expensive.

– *SLS*: This is an AM technique that uses a high power laser (e.g., a carbon dioxide laser) to fuse small particles of plastic, metal, ceramic, or glass powders into a mass that has a 3D shape. The SLS machine preheats the bulk powder material in the powder bed somewhat below its melting point. SLS is trusted by engineers and manufacturers across different industries for its ability to produce strong, functional parts.

Low cost per part, high productivity, and established materials make the technology ideal for a custom manufacturing. The material selection for SLS is limited compared to FDM and SLA.

– *FDM*: This is the most widely used form of 3DP at the consumer level, fueled by the interest of hobbyists of 3D printers. This technique is suited for basic proof-of-concept models, as well as quick and low-cost prototyping of simple parts. FDM is regarded as a very clean technology, usually simple and office-friendly. It uses a continuous filament of a thermoplastic material. The technology can produce complex geometries and cavities that would otherwise be quite problematic. Since 2004, FDM technology has been used in a particular sector to produce load-bearing scaffold. Home printer based on FDM typically works with plastic filament.

Which technology makes the most sense for you to use depends on your budget, the model's complexity, and the finest detail that is necessary. Buyers need to determine the best technology to use for a specific application. There is not a one-size-fits-all solution. The advantages of 3DP in Africa are demonstrated in Figure 9.3 [8].

Figure 9.3: The advantages of 3D printing in Africa [8].

9.4 Use of 3D Printing in Africa

3D printers employ a variety of techniques and materials and they share the ability to turn digital files containing 3D data. From manufacturing to education, the versatility of this revolutionary technology knows no bounds. The need for 3DP is expected to increase exponentially. 3DP has been applied in the following areas in Africa [9–11]:

–*Manufacturing:* At its core, 3DP is also just another manufacturing process. 3DP is becoming the chosen method of manufacturing for lots of companies. This new and versatile method of producing goods is becoming the best choice for companies in the twenty-first-century marketplace. For those organizations used to traditional production processes, 3D printers can save a tremendous amount of time. Industries like automotive, aerospace, defense, consumer goods, healthcare, apparel and fashion, and construction will stand to benefit from 3DP with reduced costs and improved lead time. The African continent has lagged behind the rest of the world in manufacturing. 3DP has the potential to revive Africa's manufacturing fortunes. Lithoz recently announced that they have installed the very first ceramic 3D printer in the African continent (Figure 9.4) [12].

– *Construction:* 3DP is one of the disruptive technologies that are significantly influencing the future of the construction industry. The Iroko, launched by 14Trees, is Africa's first commercial construction 3D printer. The Iroko printer distinguishes itself by being more affordable, reliable, and mobile than its industry counterparts. This innovation is a cornerstone in 14Trees' efforts to revolutionize building, emphasizing affordable, low-carbon housing and educational infrastructure. The deployment of Iroko 3D printer is set to dramatically enhance the speed and sustainability of constructing homes and schools across Africa. Iroko boasts many of the benefits of other concrete 3D printers, including a reduced reliance on labor and lower overall material use. This innovation will help accelerate construction of 3DP around the world. Housing crisis takes place across many countries in Africa. The shortage of classrooms in Africa is also an urgent but overlooked issue. The first-ever affordable, 3D-printed home was built in Lilongwe, the capital city of Malawi by the African construction company 14Trees. In Malawi, it has recently 3D-printed a home and a school. Figure 9.5 shows the 3DP construction of a home, while Figure 9.6 shows the first Africa's 3D printer house [13].

– *Automotive Industry:* This is among the earliest adopters of 3DP technology initially using it for making prototypes but now increasingly for the production of end-use products or parts. Automotive companies like Ford, GM, Audi, Jaguar, and Land Rover have been using 3DP for a long time now. The savings on costs and time and on the parts count are enormous. A US 3DP company, Local Motors, has made a car, christened as Strati, with a fully 3D-printed body and only with 49 parts.

– *Healthcare:* 3DP technology has the potential to revolutionize the healthcare sector by providing cost-effective solutions for medical equipment, prosthetics, and even organ transplants. With the ability to customize medical devices based on individual needs, 3DP can bridge the gap in healthcare accessibility and contribute to improving overall well-being. By providing customized medical solutions, the MedAdd project in South aims to enhance patient care and recovery times. It is a significant step toward personalized medicine in South Africa, demonstrating the potential of 3DP in addressing complex health challenges. In 2014, South African doctors used 3D-printed titanium bones to perform a jaw-bone transplant surgery, the second in the world. 3DP is also being used to make prosthetic limbs in Sudan, where the organization "Not Impossible" is helping amputees. The 3DP technology is used to create a prosthesis for the young Sierra Leonean woman in Figure 9.7 [14].

– *Education:* This is a cornerstone of progress, and 3DP can transform the learning experience. By introducing this technology into classrooms, students can engage in hands-on learning, bringing concepts to life in a tangible way. This not only enhances creativity but also prepares the future workforce with skills that align with the demands of a technology-driven world.

Figure 9.4: First ceramic 3D printer on the African continent [12].

9.5 Adapting 3D Printing in African Nations

3DP technology is the process of creating an object using a machine that extrudes the molten material layer by layer in 3D until the desired object is formed. It has gained significant momentum in recent years, expanding into a wide variety of sectors throughout the globe. By embracing 3DP, African nations can leapfrog certain stages of industrial development and bypass some of the challenges associated with traditional manufacturing. Many Africans are channeling their creativity into this technology to overcome obstacles and make changes in their communities. The cost of 3DP is

Figure 9.5: 3D printing construction of a home [13].

Figure 9.6: The first Africa's 3D printer house [13].

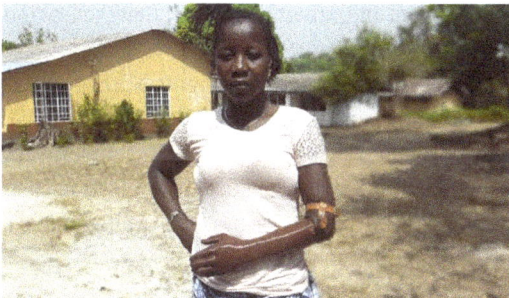

Figure 9.7: The 3D printing technology creates a prosthesis for a woman in Sierra Leone [14].

low and affordable. We consider the use of 3DP in some selected African nations [15–17]:

– *South Africa:* The MedAdd project at the Central University of Technology in South Africa focuses on the application of 3DP in the medical field. This initiative leverages AM technologies to develop medical devices and implants tailored to individual patient needs. Another project by the University of Johannesburg explores the use of 3DP technology in constructing affordable houses. The university has launched a 3DP technique that can build a house in one day. It aims to tackle the housing shortage issue in South Africa by developing rapid, cost-effective, low-cost homes. The significant research taking place in South Africa has not been replicated across the continent. In South Africa, universities are leading the drive to provide training and retraining programs for engineers and other professionals involved in using 3DP. Small enterprises are making inroads in areas such as 3DP of cell phone accessories, car accessories, and jewelry. South Africa is progressively utilizing 3DP technology to produce prosthetic devices and limbs.

– *Nigeria:* Stampar3D is a Lagos-based agency bringing 3DP technologies to the Nigerian market. Stampar claims that 3DP is a great way to take hands-on classroom learning across all disciplines to the next level. Stampar3D exemplifies the growing interest and application of 3DP in Nigerian manufacturing. 3D Printing Naija is an initiative that focuses on educating young Nigerian students about 3DP technologies. This program is considered to be one of its kind in Nigeria. As shown in Figure 9.8, *The 3D Printing Handbook* will be used as a main learning tool to train students about 3DP design [18].

– *Togo:* Originating from Togo, Woelab Lomè is an innovation hub, where the first African 3D printer, entirely made from electronic waste, was created in 2013. It is at the forefront of fostering technological innovation in Togo. Woelab Lomè not only exemplifies the creative recycling of electronic waste into valuable technology but also represents a significant step toward making 3DP accessible and affordable in underserved communities. Togo built the printer from parts, sourced from the scrapyard in Lome.

– *Gambia:* The Gambia is advancing a 3DP farm initiative focused on healthcare and education. The project aims to utilize 3DP's decentralized production capabilities, catering to the nation's minimal manufacturing infrastructure. It represents a strategic approach to overcoming local manufacturing limitations, leveraging 3DP for educational and healthcare advancements. Young entrepreneurs such as the Make 3D company are utilizing 3DP to manufacture devices and equipment for healthcare, education, and soap making purposes. 3DP makes assistive tools and other forms of 3D-printed materials more accessible and cheaper for the Gambian population.

– *Kenya:* The African Centre for Technology Studies and Kenyatta University have entered into a partnership to establish a pan-African Centre for Excellence on 3D Printing. In 2013, a prototype for the first 3D-Printed Automatic Weather Station (3D-PAWS) was developed, giving local meteorological organizations the potential to fabricate and maintain weather stations independently and at low costs. Scientists with the Kenya Meteorological Department can now sustainably fabricate, assemble, and repair weather stations using the 3D-PAWS model. Following a training workshop in June 2023, Kenya became the first FEWS NET country to adopt the model. The new 3D-printed weather stations transmit data every 15 min with observations covering rainfall, temperature, pressure, and relative humidity. In Kenya, the hardware startup AB3D aims at empowering African producers with transformative 3DP technology. AB3D is Nairobi's one-stop-shop for all things 3DP, offering access to 3D printers, 3D-printed products, workshops, and open-source designs.

– *Malawi:* Malawi alone needs to build more than 40,000 classrooms to meet the needs of the current population – something that would take more than 70 years if it relied on conventional construction methods. If 3DP methods were adopted, the classrooms could be completely cleared in just a decade. As mentioned earlier, the first-ever affordable, 3D-printed home was built in Lilongwe, the capital city of Malawi by the African construction company 14Trees. The printer operators are Malawians that were trained in Malawi to manage the activity. The buildings that 14Trees is printing are made by a computer, meaning that none of the materials is wasted and mistakes are eliminated, significantly reducing their CO_2 footprint. Children in Salima, central Malawi, have started their education at a new 3D-printed school that was built in just 15 h. The new 3D-printed school is shown in Figure 9.9 [19].

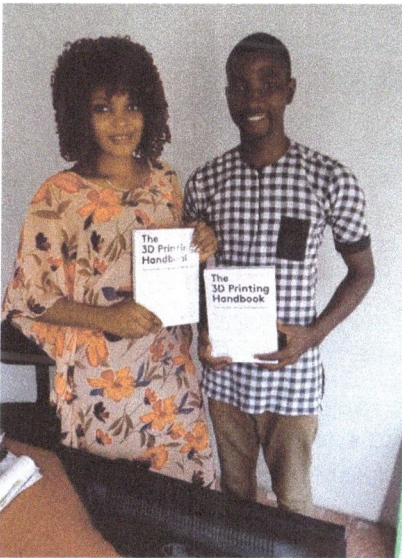

Figure 9.8: The 3D Printing Handbook is for training students about 3D printing in Nigeria [18].

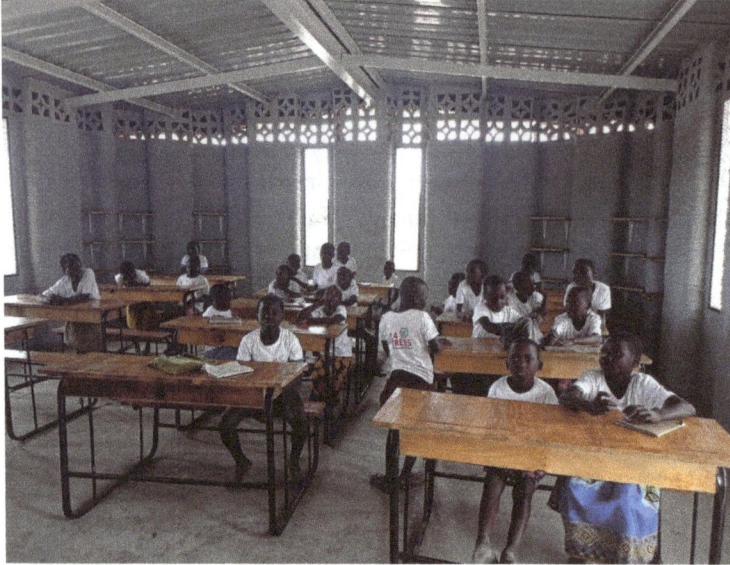

Figure 9.9: The 3D-printed school in Salima, Malawi [19].

9.6 Benefits

3DP, also known as AM, is a fabrication process in which a 3D object is built by adding layer upon layer of materials to a series of shapes. It helps to transform the way products are designed, developed, and manufactured in Africa. It is making production more sustainable by reducing transportation and material wastes. It can facilitate and enable research and development (R&D), RP, rapid tooling, rapid manufacturing, and digital mass customization in Africa. Companies are currently working on improving the affordability of this method so it can become more widely used. Other main benefits of 3DP in Africa include [20, 21]:

– *RP:* This is the use of 3D printers in the design process to create prototypes. 3DP can manufacture or prepare housing part within hours, which speeds up the prototyping process.

– *Affordable Manufacturing:* 3DP is described as the next industrial revolution bringing manufacturing to the home. With 3DP, manufacturing becomes local, affordable, and accessible. There is no more waiting for imports or relying on distant factories. We become our own manufacturers.

– *Print on Demand*: The 3D design files are all stored in a virtual library as they are printed using a 3D model as a CAD. This means they can be located and printed when needed.

– *Flexible Design:* The technology can produce very complex shapes that would be otherwise impossible to construct by hand.

– *Customization:* 3DP personalizes products according to individual needs and requirements. It enables manufacturers to create bespoke products tailored to individual customer needs, providing greater flexibility and enhancing customer satisfaction. We can design customized medical equipment and even prosthetics.

– *Preserving Cultural Heritage:* Africa boasts a rich cultural heritage, and 3DP can aid in preserving and promoting this diversity. 3DP offers a unique way to celebrate and protect Africa's cultural legacy.

– *Food Insecurity:* This problem has existed for many years in Africa. It has not only affected the health of the populations across the continent, but it has subsequently created economic hardship for millions of Africans. 3DP technology can be used to combat malnutrition in Africa by producing high-energy edible food supplements. It is very creative and leverages the technology of on-demand printing of food.

– *Scale Economies:* Big multinational manufacturing corporations invest heavily in machinery and human resources for mass production. They make big profits only if they sell enough units. The more they sell, the bigger their profit margins. 3DP does not need high-volume production and it pays to produce fewer units. All this shifts the advantage in favor of small and medium-scale enterprises.

– *Sustainability:* This is another benefit of 3DP. The process produces only what is needed and it can reuse waste materials. By using only the required material for each part, AM significantly reduces material wastes, contributing to more sustainable and cost-effective production processes. It saves on resources and reduces the cost of the materials being used.

– *Environmentally Friendly:* Production of parts only requires the materials needed for the part itself, with little or no wastage as compared to alternative methods. The environmental benefits are extended through a reduced carbon footprint. Students in Tanzania have created a 3D printer using waste electronics.

– *Government:* Government can play in the market and in knowledge production. In South Africa, the government has the most detailed policy document of any African country on 3DP. Government should support the development of 3DP technology for economic development. Smart governments are supporting skills training and innovation, developing complementary competencies, and diversifying their industries and markets.

– *Clean Drinking Water:* Access to clean drinking water remains one of the most significant challenges in African nations. Through 3DP, water filters can be crafted using locally available materials at a fraction of the traditional cost. Depending on its origin and usage, such a water filter can provide up to 4 years of clean drinking water, making a substantial impact on improving access to this essential resource.

– *Water Pumps:* With a 3D printer and various components printed by local workshops, water pumps for wells can be constructed, guaranteeing the extraction of clean fresh water from the ground even during drought periods. This innovation could be a lifeline to numerous communities and has the potential to save lives. 3D-printed wells installed in underdeveloped communities would not only address immediate water needs but also showcase the power of community collaboration. It illustrates how 3DP empowering entire communities to leverage this technology for sustainable solutions.

– *Transportation:* Transportation stands as a pressing priority for any underdeveloped nation. In certain African countries, bicycles are frequently utilized as primary modes of transportation due to their ability to navigate a wide variety of terrains, as shown in Figure 9.10 [21]. Leveraging 3DP technology, bicycles can be relatively easily produced in regions where parts may be scarce. 3D printers could be deployed to create the whole bicycle.

– *Housing:* The housing crisis has hit developed countries hard in recent years. In remote places, building homes and infrastructure is no small feat. 3DP holds immense potential for widespread adoption as a swift alternative for constructing temporary homes for families displaced by natural disasters, disease outbreaks, or conflict. 3DP can create houses, parts for infrastructure, and even water systems. This offers a sustainable solution for addressing housing needs. Dreams of better living conditions are within reach.

Figure 9.10: Bicycles are the primary modes of transportation [21].

9.7 Challenges

Just like any other technology, 3DP has its downsides. In spite of its many benefits, AM still faces technical challenges and limitations, such as size constraints, material compatibility, material selection, and post-processing requirements. Some people are concerned about whether 3DP will destroy or build the world. Some feel the technology will infringe on intellectual property. Other challenges of 3DP in Africa include the following [22, 23]:

– *Infrastructure:* Traditional manufacturing processes often require extensive infrastructure, which can be a hurdle for many African nations. 3DP offers a decentralized solution that can operate efficiently even in areas with limited infrastructure.

– *High Cost:* Although 3DP is already a game changer in the manufacturing industry, there is great unmet need in developing nations where the sector has been hindered by high costs. The cost of 3DP can depend on several factors such as the type of printer, quality of prints, materials used, and design complexity. The cost of 3DP can be a significant factor that can limit its widespread adoption, especially for small businesses and individuals. Some worry that 3DP may not be the cost-saving solution it is being touted as.

– *Affordability:* Some in Africa often resort to the reasoning that emerging technologies such as 3DP, though revolutionary and beneficial, are unaffordable for them and have to wait for years before they can adopt them. The adoption lag of technologies has drastically shrunk in recent times. Companies are working on making 3DP more portable and affordable.

– *Jobs at Risk:* 3D technology reduces the need for human labor, since most of the production is automated and done by printers. This technology could put construction jobs at risk by significantly reducing the need for human labor. Oxford University states that 85% of Ethiopian jobs are at risk of being replaced by 3DP. Some industries are more affected than others.

– *Environment:* Environmental destruction is caused by resource extraction for concrete from construction projects.

– *Threat:* The biggest advantage of 3DP is also its biggest threat, that is, the affordability and accessibility of 3DP. The technology brings with it a lot of uncertainty around intellectual property protection, as the ability to make unauthorized copies is becoming much easier. Patent-infringing and counterfeit goods are only going to get easier to make and distribute. All types of intellectual property, copyright, trademarks, registered designs, and patents could be infringed and affected by 3DP.

– *Material Selection:* Most 3D printers are designed to process either polymers, metals, composites, ceramics, or glass. Material selection is challenging because not all of the materials that can be used in production are available for 3DP. Some metals and polymers cannot be temperature-controlled enough to support AM. Material selection can be difficult enough because of the trade-offs involved in balancing an application's requirements against its material's properties.

– *Repeatability:* This is also a challenge for the 3DP industry, that is, because the location of the build on the printing surface can affect the height, width, depth, and weight of the final product. This lack of repeatability can reduce yields and slow throughputs.

9.8 Conclusion

3DP is a process of making 3D solid objects from a digital file using additive processes. Adopting 3DP can strengthen the manufacturing industry in Africa. Pioneering African entrepreneurs have embraced the 3D technology, using it to develop malaria testing devices in Zambia, and agricultural tools that farmers across the continent can print out themselves. Embracing 3DP is not just a dispensable luxury but an indispensable necessity for Africa's growth and development. 3DP can be a game changer for small and medium-sized enterprises in Africa. It can potentially be grown to solve much more significant challenges such as housing and settlement. Because most African nations generally lack essential technology and industry skills, there is no near-term scenario under which they would be able to leverage like 3DP to compete globally in manufacturing.

Although the progress of 3DP is slow and invisible due to multiple obstinate challenges, for African countries rich in natural resources, a significant opportunity may lie in supplying and producing metals for metal 3DP systems. African nations should invest in infrastructure, educated personnel, industry professionals, R&D, and innovation institutions to enhance the growth rate of 3DP technology.

Intra-African trade may hold promise for 3DP across the continent. Once the technology becomes accessible on a large scale, the possibilities seem endless.

References

[1]   M. N. O. Sadiku, P. O. Adebo, and J. O. Sadiku, "3D printing in Africa," *International Journal of Trend in Research and Development,* vol. 11, no. 3, May-June 2024, pp. 139–145.

[2]   F. R. Ishengoma and T. A. B. Mtaho, "3D printing: Developing countries perspectives computer engineering and applications," *International Journal of Computer Applications,* vol. 104, no. 11, October 2014, pp. 30–34.

[3]   M. N. O. Sadiku, S. M. Musa, and O. S. Musa, "3D Printing in the chemical industry," *Invention Journal of Research Technology in Engineering and Management,* vol. 2, no. 2, February 2018, pp. 24–26.

[4]   D. Pitukcharoen, "3D printing booklet for beginners," https://www.metmuseum.org/-/media/files/blogs/digital-media/3dprintingbookletforbeginners.pdf

[5] "The benefits of 3D Printing in children's education," February 2021, https://www.kidsinthehouse.com/blogs/kidsinthehouse2/the-benefits-of-3d-printing-in-childrens-education

[6] "Introduction to 3D printing," https://education.gov.mt/en/resources/news/documents/youth%20guarantee/3d%20printing.pdf

[7] "Guide to 3D printing materials: Types, applications, and properties," https://formlabs.com/blog/3d-printing-materials/

[8] "Benefits of additive manufacturing (3D printing) for Africa," https://www.3dprint.africa/

[9] W. Worku, "The crucial role of 3d printing technology for Africa's future," January 2024, https://www.linkedin.com/pulse/crucial-role-3d-printing-technology-africas-future-wogayehu-worku-vm1pe#:~:text=3D%20printing%20offers%20a%20decentralized,challenges%20associated%20with%20traditional%20manufacturing

[10] M. Molitch-Hou, "Africa's first construction 3D printer developed via Holcim's 14Trees," July 2023, https://3dprint.com/301847/africas-first-construction-3d-printer-developed-via-holcims-14trees/

[11] S. Kolade, "3D printing offers African countries an advantage in manufacturing," April 2022, https://theconversation.com/3d-printing-offers-african-countries-an-advantage-in-manufacturing-179777

[12] K. Stevenson, "African 3D printing grows," April 2022, https://www.fabbaloo.com/news/african-3d-printing-grows

[13] "The affordable 3D-printed home that could transform African urbanization," June 2021, https://www.shareyourgreendesign.com/the-affordable-3d-printed-home-that-could-transform-african-urbanization/

[14] "Affordable prosthetics in Africa made possible by 3D printing," September 2019, https://www.utwente.nl/en/news/2019/9/523070/affordable-prosthetics-in-africa-made-possible-by-3d-printing

[15] "3-D printing: Africa's most exciting projects," https://africalive.net/article/3-d-printing-africas-most-exciting-projects/

[17] L. Barnard, "3D printing: New tech being used to build in Africa," June 2022, https://www.constructionbriefing.com/news/3d-printing-new-tech-being-used-to-build-in-africa/8021557.article

[18] "Supporting 3D printing education in Nigeria with The 3D Printing Handbook," https://www.hubs.com/blog/supporting-3d-printing-education-in-nigeria/

[19] "Could 3D printed schools be 'transformative' for education in Africa?" July 2021, https://www.weforum.org/agenda/2021/07/could-3d-printed-schools-be-transformative-for-education-in-africa/

[20] B. W. Mwangi, "The art that is 3D printing," February 2022, https://housingfinanceafrica.org/documents/the-art-that-is-3d-printing/

[21] "How can 3D printing benefit underdeveloped countries?" https://amfg.ai/2024/03/14/how-can-3d-printing-benefit-underdeveloped-countries/#:~:text=In%20underdeveloped%20communities%2C%203D%20printers,planet's%20global%20plastic%20pollution%20crisis.

[22] B. Ndemo, "3D Printing and Africa's manufacturing renaissance," https://www.acts-net.org/blogs/foresight-africa-blog/3d-printing-and-africa-s-manufacturing-renaissance#:~:text=The%20application%20of%203D%20printing,ICT)%20boom%20is%20rapidly%20spreading.

[23] L. Galloway, " What are the biggest challenges for the 3D printing industry?" March 2022, https://bmf3d.com/blog/3d-printing-industry-challenges/

Chapter 10
Nanotechnology in Africa

While technology lowers the number of repetitive and physically intense jobs, it creates others that didn't exist before. This is particularly true in the area of nanotechnology, an emerging technology that is already transforming our world. – Maria Fonseca

10.1 Introduction

Africa is closely watched as the next big growth market. It is the home to some of the youngest populations in the world and also to many fast-growing economies. Africa is a booming continent with incredible growth potential, as the second-largest continent in the world and the world's largest free trade area, connecting 1.3 billion people (16.6% of the world population) across 55 nations. Africa, a continent rich in resources and diverse cultures, has the potential to harness nanotechnology to uplift local industries. Africa has a large genetic diversity with varying profiles from north to south and from east to west.

The science and technology of manipulating a matter at the nanoscale (to the level of atoms and molecules) presents us with nanotechnology. Nanotechnology is the manipulation of matter on a near-atomic scale to produce new structures, materials, and devices. Since an atom is the smallest particle and nanotechnology empowers us to control atoms, nanotechnology is considered the biggest achievement ever. Nanotechnology is one of the most exciting and fast-moving areas of science and engineering today.

Nanotechnology has been reported as the new industrial revolution. The past decade has proven the applicability of nanotechnology in almost all fields. Nanotechnology has been predicted to be a main driver of technology and business in this century. It has become the new frontier of science and technology around the world. Both developed and developing nations are investing in nanotechnology to secure a market share. The global demand for nanotechnology in all industries is evidenced by the ever-growing volume of investment made by the private investors. Nanotechnology promises significant improvements of advanced materials and manufacturing techniques, which play a critical role for the future industries. Breakthroughs in nanotechnology provide new opportunities to commercialize novel materials at the nanoscale. Nanotechnology is poised to impact dramatically on all sectors of industry [1, 2].

Nanotechnology is the manipulation, production, and characterization of structures that have one dimension smaller than 100 nm. It is a disruptive emerging technology dedicated to the study and manipulation of characteristics of matter at the atomic level. It is a broad general-purpose technology with applications in any field imaginable. Over the years, the science and applications of nanotechnology have de-

https://doi.org/10.1515/9783112211984-010

veloped at a faster pace across various disciplines and sectors in various nations. All around the world, nanotechnology is being promoted as a technological revolution that will help solve an array of problems. The African Union recognizes nanotechnology as a compelling imperative and identifies nanotechnology as one of six priority areas in its Science, Technology and Innovation Strategy for Africa 2024. In Africa, nanotechnology has already been recognized as an important tool for industrial development [3].

This chapter explores the use of nanotechnology in Africa. It begins with describing what nanotechnology is all about. It discusses the uses of nanotechnology in Africa. It covers African nations that have adapted nanotechnology. It highlights the benefits and challenges of nanotechnology in Africa. The last section concludes with comments.

10.2 What Is Nanotechnology?

Technologies impact every aspect of our modern society. There are many ways in which our society and technology are interlinked. Nanotechnology has the potential to provide huge benefits, just like any useful technology.

The term "nano" means something small, tiny, and atomic in nature. The application of the term in science led to a field called nanotechnology. Nanotechnology refers to the characterization, fabrication, and manipulation of structures, devices, or materials that have one or more dimensions that are smaller than 100 nm. It may be regarded as an area of science and engineering, where phenomena that take place at the nanoscale (10–9 m) are utilized in the design, production, and application of materials and systems. It is an emerging area that integrates chemistry, biology, and materials science to create new properties that can be exploited to gain new market opportunities.

Nanotechnology deals with the characterization, fabrication, and manipulation of biological and nonbiological structures smaller than 100 nm. Dimensions between approximately 1 and 100 nm are known as the nanoscale. As indicated in Figure 10.1 [4], the nanoscale is so small that we cannot see it with a light microscope. It is the scale of atoms and molecules. Nanotechnology involves the creation and application of materials and devices at the level of molecules and atoms. It may be regarded as the science that is conducted, researched, investigated, and experimented at the nanoscale. Nanotechnology is a multidisciplinary field that includes biology, chemistry, physics, materials science, and engineering. It is the science of small things – at the atomic level or nanoscale level. The past three decades have witnessed an increased interest and funding in nanotechnology. This has led to rapid developments in all areas of science and engineering [5].

The Scale of Things – Nanometers and More

Things Natural   **Things Manmade**

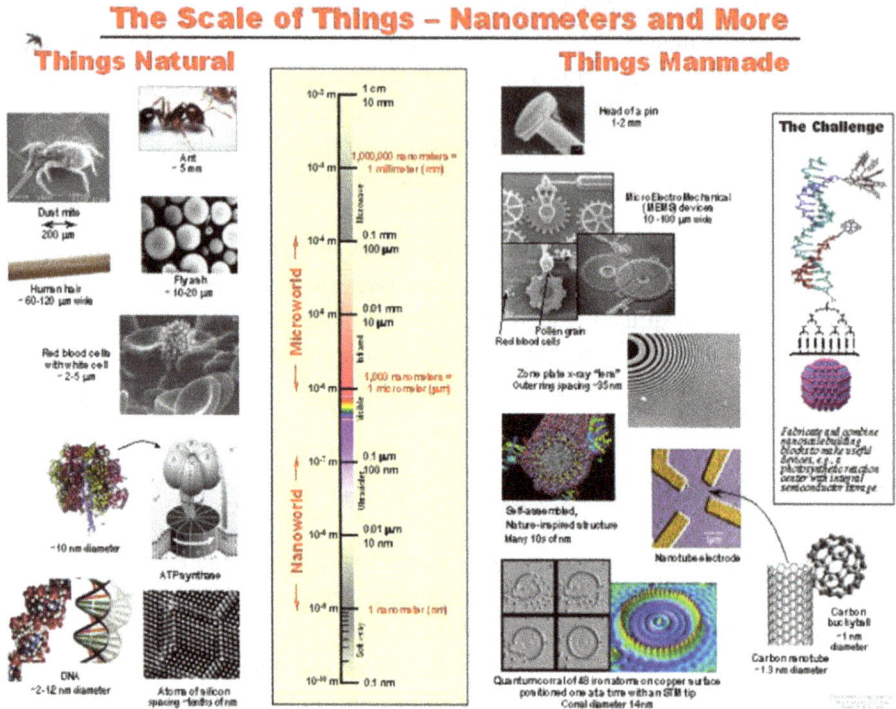

Figure 10.1: The relative scale of nanosized objects [4].

Richard Feynman, the Nobel Prize-winning physicist, introduced the world to nanotechnology in 1959 and is regarded as the father of nanotechnology. Nanotechnology involves the manipulation of atoms and molecules at the nanoscale so that materials have new unique properties. Nanomaterials are expected to have at least one dimension (length, width, and height) at the nanoscale of 1–100 nm. One nanometer is a billionth of a meter, too small to be seen with a conventional lab microscope. Nanomaterials are known as nanoparticles when they have nanoscale length, width, and height.

Today, nanotechnology is part of our daily lives. Nanotechnology will leave virtually no aspect of life untouched. Its usages include everything from safer food processing to more efficient drug delivery systems to powerful computer chips. Three steps to achieving nanotechnology-produced goods are [6]:

1. Scientists must be able to manipulate individual atoms.
2. Next step is to develop nanoscopic machines, called assemblers, that can be programmed to manipulate atoms and molecules at will.
3. In order to create enough assemblers to build consumer goods, some nanomachines called replicators will be programmed to build more assemblers.

Nanotechnology is trending among scientists and engineers. Here are some underlying trends one should look for [7]:
1. Stronger Materials: The next generation of graphene and carbon devices will lead to even lighter but stronger structures.
2. Scalability of Production: One big challenge is how to produce nanomaterials that make them affordable. Limited scalability often hinders application.
3. More Commercialization: In addition to transforming the automotive, aerospace, and sporting goods fields, nanotechnology is facilitating so many diverse improvements: thinner, affordable, and more durable.
4. Sustainability: One main goal of the National Nanotechnology Initiative, a US government program coordinating communication and collaboration for nanotechnology activities, is to find nanotechnology solutions to sustainability.
5. Nanomedicine: There will be a mind-boggling impact of nanotechnology on medicine, where advances are being made in both diagnostics and treatment areas.

Figure 10.2 shows a six-pronged strategy approach that Africa can use for sustainable nanotechnology innovation [8].

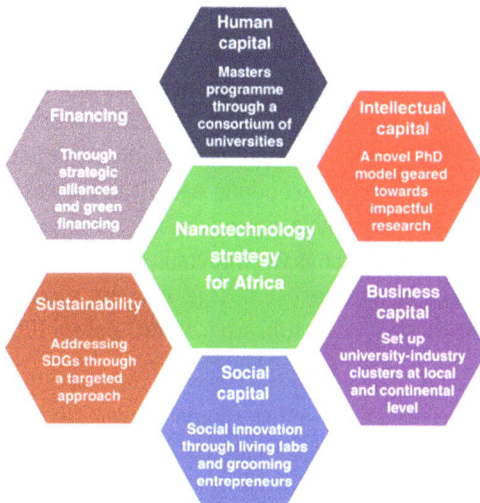

Figure 10.2: A strategy Africa can use for sustainable nanotechnology innovation [8].

10.3 Use of Nanotechnology in Africa

Nanotechnology is the study, design, creation, synthesis, manipulation, and application of functional materials, devices, and systems through control of matter at the nanometer scale. It has the potential to affect many industrial sectors and every economic sector across the African continent. Some applications of nanotechnology are shown in Figure 10.3 [9]. Major contributors to nanotechnology in Africa (1995–2011) are presented in Figure 10.4 [10]. At the moment, research centers in various African countries are developing the state-of-the-art research in nanotechnology in areas such as agriculture, energy, healthcare, and electronics:

– *Healthcare:* In the health sector, nanotechnology has contributed to the elaboration of more efficient and reliable medical devices such as diagnostic biosensors, drug delivery systems, and imaging probes. Nanomedicine is a science, which plays a crucial role in both health and medicine. Nanotechnology-based approaches address challenges in the diagnosis and treatment of noncommunicable diseases such as diabetes through the noninvasive monitoring of islet cell mass. The application of nanotechnology in medicine especially promises to offer unique potential for advances in the diagnosis, treatment, and prevention of diseases such as HIV/AIDS, tuberculosis, malaria, and Ebola. Biotechnologists along with nanotechnologists can fabricate new generation medicine or even a nanorobot programmed to target cancer cells. This will contribute to nanotechnology in healthcare. Figure 10.5 shows nanorobots, which are found to clear harmful bacteria and toxins [11]. Using nanotechnology, physicians can also deliver heat therapy used in conjunction with chemotherapy with greater precision. Nanotechnology may also improve on radiation therapy. The Council for Scientific and Industrial Research in South Africa is one of the leading scientific and technology research, development, and implementation organizations in Africa. Research on nanotechnology and nanomaterials will have a widespread impact in health, information, energy, and many other fields, where there is a major economic benefit to the commercialization of new technologies. Research will provide renewed hope for patients diagnosed with cancer. The ultimate goal of the research is to go beyond the laboratory experimental work, to answer questions such as appropriate dosage, delivery system, and exposure times. Figure 10.6 shows a researcher on healthcare nanotechnology [12].

– *Agriculture:* In the agricultural sector, African countries have to produce more and better with less. This implies the need for a more sustainable, efficient, and resilient agricultural system, while promoting food security. Engineered nanomaterials can improve the use efficiency of water, light, and agrochemicals, and can improve soil integrity.

– *Energy:* Many nations in Africa are currently experiencing a shortage of electricity to supply the residential, commercial, and industrial sectors. Research in South Africa focuses on the development of energy-saving and power-efficient materials at nanoscale in response to the tremendous increase in energy consumption. Figure 10.7 shows the drivers of nanotechnology in energy sector [10].

– *Government:* Nanotechnology can help African governments to develop innovation and create job opportunities. Scientific innovations create value by developing new products and services, providing solutions to social problems, creating new enter-

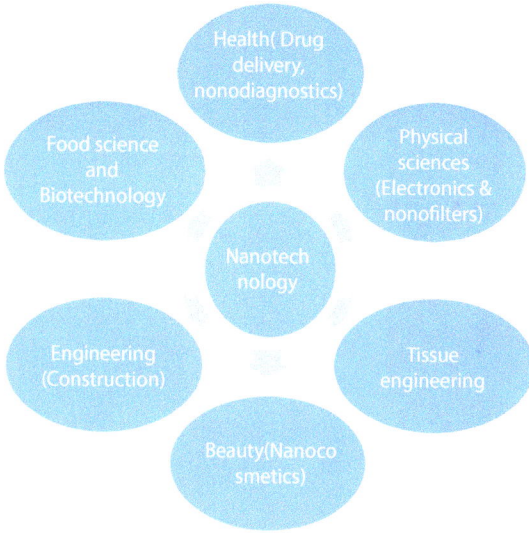

Figure 10.3: Some applications of nanotechnology [9].

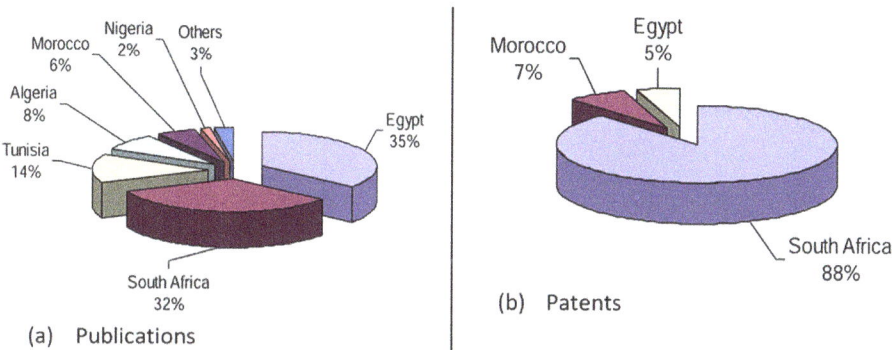

Figure 10.4: Major contributors to nanotechnology in Africa (1995–2011) [10].

prises and jobs, thereby improving the quality of life. The government reluctantly took up the governance of risk after being faced with increasing pressure by civil society, scientists, and industry.

10.4 Adapting Nanotechnology in African Nations

Egypt, South Africa, Tunisia, Nigeria, and Algeria lead the field of nanotechnology in Africa. Egypt, Nigeria, and South Africa are making progress in nanotechnology research and applications, but a lot needs to be done for Africa to become a global

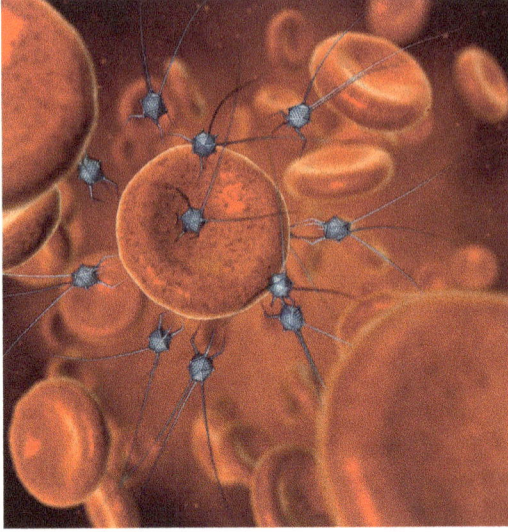

Figure 10.5: Nanorobots are found to clear harmful bacteria and toxins [11].

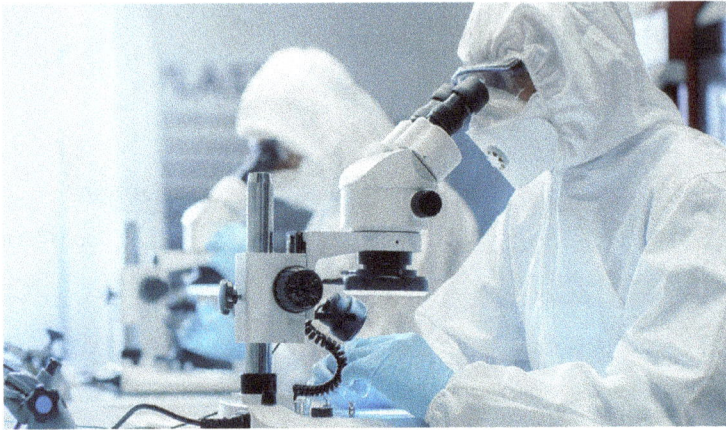

Figure 10.6: A researcher on healthcare nanotechnology [12].

player in the field. We consider the use of nanotechnology in some selected African nations:

– *South Africa:* The population of the country was 60.6 million as of 2022. It has abundant natural resources such as gold, gemstones, diamond, platinum, coal, phosphates, rare earth elements, uranium, salt, natural gas, and several other metals. South Africa is one of the first countries to have an official nanotechnology strategy. It is one of the few countries from the global South that have adopted nanotechnology with the aim

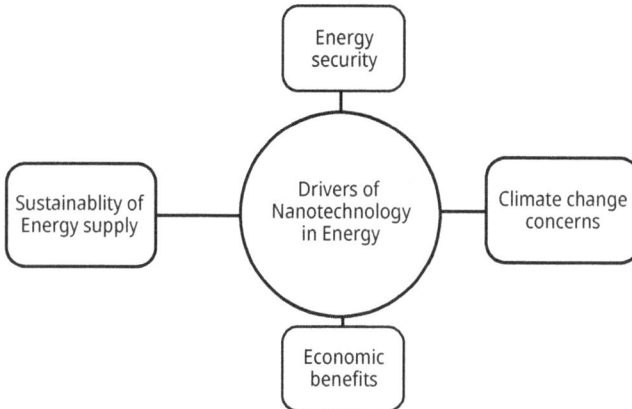

Figure 10.7: Drivers of nanotechnology in energy sector [10].

of enhancing global competitiveness and sustainable economic growth. Since 2006, South Africa has been developing scientists, providing infrastructure, establishing centers of excellence, developing national policy, and setting regulatory standards for nanotechnology. The world's five major emerging economies, namely Brazil, Russia, India, China, and South Africa, form a grouping named the BRICS. Compared to other BRICS nations, South Africa has the lowest nanotechnology productivity. However, nanotechnology in South Africa has grown exponentially, publications per year increased from 68 in 2000 to 1,672 in 2019. In 2019, India was the biggest collaborating partner for South Africa, while there was increasing inter-Africa collaboration in nanotechnology [13]. Today, almost all public universities and science councils in South Africa are working on nanotechnology research and incorporating nanomaterials. The strategic socioeconomic-focused nanotechnology research areas identified for South Africa include materials science, optics, medicine, catalysis, electronics, energy, biotech, magnetism, sensors, water, and communicable diseases. The progress of nanotechnology in South Africa is illustrated in Figure 10.8 [14]. Most of South Africa's roads have reached the end of their 20-year design life. The government is piloting the use of "nanotechnology" in an effort to address the country's crumbling road infrastructure. The poor road infrastructure has had an incalculable toll on the country's productivity, threatening supply chains. The government is using nanotechnology to fix millions of potholes, as typically shown in Figure 10.9 [15].

– *Nigeria:* Nanotechnology has much to offer Nigeria but research needs support. Nigeria's nanotechnology journey, which started with a national initiative in 2006, has been slow. It has been dogged by uncertainties, poor funding, and lack of proper coordination. Nigerian nanotechnology research is limited by a lack of sophisticated instruments for analysis. In the absence of state budget allocations and a national policy on nanotechnology research, Nigeria is missing out on the potential benefits of this field. Nigeria's engagement with nanotechnology is limited to research at some universities and is hindered by a lack of advanced equipment to study nanomaterials. University of Nigeria, Nsukka, and Ladoke Akintola University of Technology, Ogbomoso, are the two institutions that are contributing to knowledge in nanotechnology research among public-funded universities in Nigeria. South Africa, Malaysia, India, the United States, and China are the main collaborators with Nigeria in the field. Standards do not exist for nano-based products [16, 17]. Efforts are being made by the academia, research institutes, and government to create awareness and interest in nanotechnology development in Nigeria. The Nigeria Nanotechnology Initiative has identified priority areas in medicine, agriculture, energy, and water purification that can be met by developing strong capabilities in nanotechnology.

– *Ghana:* In its quest to accelerate improvement in agricultural productivity, the government of Ghana is already exploring the use of biotechnology products in collaboration with the country's research institutions. At the First Pan-African Biotechnology Stewardship Conference in Accra in 2011, modern biotechnology tools were seen to have considerable promise to develop crop varieties and livestock breeds to withstand the most adverse conditions in the country. Ghana seeks to reap the full benefits of biotechnology through the modernization and commercialization of agriculture without sacrificing biosafety concerns [18].

– *Egypt:* This nation is currently the top nanotechnology research country in Africa, while South Africa is the African country that has filed the most patents and established the most nanotechnology companies and institutions. The Egypt Nanotechnology Center (EGNC) and IBM partnership are examples of a successful public and private collaboration in nanotechnology research.

– *Uganda:* Although there are products of nanotechnology in public use in Uganda, the popularity is still very low. There are several concerns about the lack of adequate policy, legal, and institutional frameworks to effectively govern nanotechnology and its products.

– *Kenya:* A large portion of Kenya's population lives below the poverty line. While Kenya ranks third in sub-Saharan Africa in terms of the number of scientific publications (after South Africa and Nigeria), its science base remains small. Kenya has only recently been active in nanotechnology. The government of Kenya first spoke of nanotechnology in 2009. Kenya has attempted to make nanotechnology relevant for development, freeing Kenya from their dependence of developed countries. The Kenyan coproduction of nanotechnology and development largely remains a discursive exercise. There have been no concerted efforts to fund nanotechnology research, either by government or industry. In Kenya, nanotechnology was coproduced with development by envisioning that nanotechnology's diverse and extensive benefits could bring development by freeing Kenya from foreign dependencies [19].

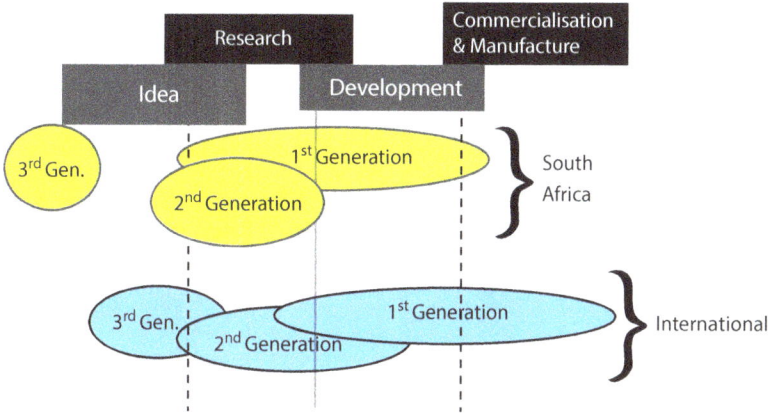

Figure 10.8: The progress of nanotechnology in South Africa [14].

Figure 10.9: Nanotechnology is fixing millions of potholes [15].

10.5 Benefits

Like any technology, nanotechnology creates value by developing new products and services, solving social problems, creating new enterprises, and generating employment opportunities. Nanotechnology has a wide range of socioeconomic benefits and significantly contributes toward the achievement of the Sustainable Development Goals (SDGs). Its application can generate various benefits, such as health, energy, water, chemical and bioprocessing, mining and minerals, and advanced materials. The multidisciplinary nature of nanotechnology and its vast applications offer many opportunities to African students. The use of nanomaterials has led to improvements in production, packaging, shelf life, and bioavailability. Other benefits of nanotechnology in Africa include the following [20, 21]:

– *Changes on the Cellular Level:* Nanotechnology has the potential of restructuring items at a cellular level.

– *Extending Human Life:* There is the potential to cure difficult diseases such as cancer and slow down or possibly even stop the aging process.

– *Creating Self-Repairing Technology:* The nanotechnology innovation could be used to virtually self-repair anything.

– *Alleviating Poverty:* Nanotechnology could create many high-paying new jobs, children born with addictions and diseases could be cured, and it has the potential to allow everyone equal access to many opportunities.

– *Security:* Nanomaterials in gadgets and vehicles can enhance protection and capabilities of personnel.

– *Environment:* Nanotechnology is inherently green and environmental friendly. It has so many applications that could help the environment. We will even be able to cure the damages of the environment brought by previous technologies.

– *Water Problem:* The successful treatment of the groundwater at Madibogo, a village in South Africa, convinced the community to accept nanotechnology as a solution to water problems. The implementation of a nanotechnology for improving water quality depends on infrastructure and capacity development.

10.6 Challenges

Like any technology, nanotechnology has its own challenges. Challenges include the processing of domestication and implementing the law and the rapid advancements in the pharmaceutical sector in general and nanotechnology in particular. One challenge is risk colonization of nanotechnology in developing nations. Other challenges of nanotechnology in Africa include the following [20–25]:

– *Weaponized:* If cellular repair is possible, then so can cellular destruction. It is possible that nanotechnology could be weaponized to create delivery systems that could dispose of a population while leaving its infrastructure intact.

– *Making Current Energy Technologies Obsolete:* Numerous sectors of industries are built on fossil fuels. Nanotechnology could make these technologies obsolete.

– *Creating New Diseases:* There is no guarantee that the problems nanotechnology could solve would not just generate new problems without solutions in the future.

– *Creating a New System of Class Identity:* Nanotechnology may ultimately provide low-cost food options, but there is always the possibility that one nation or group will hoard this technology.

– *Collaboration:* International and local collaborations are important for the development of nanotechnology. Internationally, India is the largest collaborating partner for South Africa, while in Africa, Nigeria is the most significant collaborator. To increase relevant nanotechnology research in Africa, universities and various scholars should embrace collaboration. There is an urgent need to create a critical mass of specialists in this area through collaboration, advanced training, retraining, conferences, short courses, and workshops.

– *High Cost:* In 2011, South Africa became the first nation in the African continent to own a new US$15 million electron microscopy center. There is no way students will comprehensively understand the nature of nanomaterials without adequate equipment like electron microscopes. However, cutting-edge equipment for nanotechnology characterization available in research labs in South Africa is expensive to maintain and quickly becomes outdated, making it difficult for scientists in South Africa to keep up with scientists from around the world.

– *Commercialization:* The commercialization of nanotechnology is still a challenge in Africa. Only a few African countries (such as South Africa, Egypt, and Morocco) are currently showing an interest in nanotechnology commercialization. For Africa to effectively advance nanotechnology commercialization, domestic and international investment is required. The success in nanotechnology commercialization will be determined by nano-enabled product quality as well as by accessibility and affordability.

– *Potential Risks:* The risks of nanotechnology are not really the risks of a technology, but of the applications. There may be some applications of nanotechnology, which can be hazardous and pose danger. The dual-use nature of some nanotechnologies poses security risks. Unlike human exposure, the number of species potentially at risk from nanoparticles is extremely large. We need to be particularly vigilant of the ethics and dangers of it. Governments have identified potential risks of nanotechnology to human health and the environment as an important issue for governance. At the nanoscale, materials gain a number of new properties that may also give rise to new risk properties. The spread of nanotechnology around the world confirms the claim that we live in a world risk society. We should not be imagining a risk-free world.

– *Poor Infrastructure:* The research infrastructure for nanotechnology is underdeveloped. There are not enough skilled researchers. Standards for nanotechnology products are lacking.

– *Complexity:* Nanotechnology is characterized by complex and heterogeneous cycles of hope, expectation, hype, and disappointment.

– *Skills Gap:* General nanotechnology development will require specialized skills in various fields such as medicine and pharmacy. The continent of Africa has skills and human capacity gaps to apply nanotechnology. The current landscape of education and training includes a number of stakeholders, with traditional universities being the most notable. There is a need for deliberate efforts by training institutions to actively involve local and international industry.

– *Resource Constraints:* Resource constraints on the African continent can be handled by the establishment of regional centers of excellence in nanotechnology with agreed governance structures that are financed by the private sector, the public sector, and international organizations.

– *Intellectual Property Concerns:* Nanotech advancements often lead to the creation of novel intellectual property. Negotiating equitable sharing arrangements and respecting intellectual property rights may become complex issues in international collaborations.

– *Ethical Considerations:* As nanotechnology evolves, ethical concerns surrounding its applications, such as in human enhancement or surveillance, will necessitate diplomatic discussions.

– *Technology Transfer Barriers:* Developing nations may face challenges in accessing nanotechnology due to barriers in technology transfer. Diplomatic efforts are hence required to ensure a fair and inclusive distribution of nanotech benefits.

10.7 Conclusion

Nanotechnology is one of the engines of the Fourth Industrial Revolution. It creates, uses, and studies materials at nanoscale. It is sweeping the world, including Africa. It is a new multidisciplinary field that holds the potential to revolutionize the global economy in the twenty-first century, with its footing in chemistry, physics, molecular biology, and engineering.

Africa is lagging behind other continents in terms of nanotechnology research, inventions, standards, and the number of companies operating in that area. The majority of African nations is still in the early stages of R&D or has not prioritized nanotechnology. Africa is at risk of becoming further marginalized in technology development and/or its governance. African countries cannot invest in nanotechnology across the board, but they have to focus on critical strategic nanotechnology research areas that possess the most significant potential to bring socioeconomic development. Developing nanotechnology has been articulated as being beneficial for development by enabling developing countries to free themselves from their dependence on the world's superpowers. Ultimately, nanotechnology will profoundly affect Africa's economy, regardless of its level of direct participation. More information about nanotechnology and nanomaterials can be found in the books in [26–30] and the following related journals/magazines:

- *Nanotechnology*
- *Nanoscale*
- *Nano: The Magazine for Small Science*
- *Micro and Nano Technologies*
- *Nanotechnology News*
- *Nature Nanotechnology*
- *Current Research in Nanotechnology*
- *American Journal of Nanotechnology & Nanomedicine*
- *Nanomedicine: Nanotechnology, Biology and Medicine*
- *Journal of Nanotechnology*
- *Journal of Nanoparticle Research*
- *Journal of Bioelectronics and Nanotechnology*
- *Journal of Nanoscience and Nanotechnology,*
- *Journal of Micro and Nano-Manufacturing*
- *Journal of Nanoengineering and Nanomanufacturing*
- *Nanotechnology and Precision Engineering*
- *South African Journal of Science*

References

[1]   M. N. O. Sadiku, U. C. Chukwu, A. Ajayi-Majebi, and S. M. Musa, "Nanotechnology in industry," *Journal of Scientific and Engineering Research*," vol. 8, no. 4, 2021, pp. 106–111.

[2]   M. N. O. Sadiku, M. Tembely, and S. M. Musa," Nanotechnology: An introduction," *International Journal of Software and Hardware Research in Engineering*, vol. 4, no. 5, May. 2016, pp. 40–44.

[3]   M. N. O. Sadiku, P. O. Adebo, and J. O. Sadiku, "Nanotechnology in Africa," *International Journal of Trend in Research and Development*, vol. 11, no. 3, May-June 2024, pp. 146–151.

[4]   "Nanotechnology white paper," https://www.epa.gov/sites/default/files/201501/documents/nano technology_whitepaper.pdf

[5]   M. N. O. Sadiku, Y. P. Akhare, A. Ajayi-Majebi, and S. M. Musa, "Nanomaterials: A primer," *International Journal of Advances in Scientific Research and Engineering*, vol. 7, no. 3, March 2020, pp. 1–6.

[6]   K. R. Saravana and R. Vijayalakshmi, "Nanotechnology in dentistry," *Indian Journal of Dental Research*, November 2005.

[7]   N. S. Giges, "Top 5 trends in nanotechnology," March 2013, https://www.asme.org/topics-resources /content/top-5-trends-in-nanotechnology

[8]   D. Jhurry, "Can Africa risk missing the nanotechnology revolution?" January 2022, https://www.uni versityworldnews.com/post.php?story=20220116084516992

[9]   D. E. Effiong et al., "Nanotechnology in cosmetics: Basics, current trends and safety concerns – A review," *Advances in Nanoparticles*, vol. 9, 2020, pp. 1–22.

[10] P. U. Akpan, "Nanotechnology status in Africa (1995–2011): A scientometric assessment," https://nanotechunn.com/proceedings/Nanocon314-24%20Akpan%20-%20Nanotechnology%20Sta tus%20in%20Africa.pdf

[11] L. Scovian, "Nanotechnology research increases significantly," August 2021, https://www.nature.com/articles/d44148-021-00069-2

[12] "Nanotechnology in healthcare," August 2019, Unknown Source.

[13] B. Masara, J. A. Van der poll, and M. Maaza, "A nanotechnology-foresight perspective of South Africa," *Journal of Nanoparticle Research*, vol. 23, no. 4, 2021.

[14] https://www.researchgate.net/figure/Progress-of-nanotechnology-in-South-Africa-related-to-the-world-South-African-Department_fig4_315045228

[15] "South African government to use "nanotechnology" to fix millions of potholes," July 2023, https://topauto.co.za/news/82919/south-african-government-to-use-nanotechnology-to-fix-millions-of-potholes/

[16] L. Scovian, "Nigeria is missing out on nanotechnology opportunities," April 2022, https://www.universityworldnews.com/post.php?story=20220418224009330#:~:text=In%20the%20absence%20of%20state,according%20to%20a%20new%20study.

[17] L. Agbaje, "Nanotechnology has much to offer Nigeria but research needs support," April 2022, https://www.lautech.edu.ng/news/nanotechnology-has-much-offer-nigeria-research-needs-support

[18] J. K. Kiplagat, "Nanotechnology: Key to sustainable development in Africa," *Proceedings of the Sustainable Research and Innovation Conference*, Kenya, October, 2020, pp. 160–164.

[19] K. Beumer, "The co-production of nanotechnology and development in India, South Africa, and Kenya," in D. M. Bowman et al. (eds.), *Practices of Innovation and Responsibility – Insights from Methods, Governance and Action*. IOS press, 2015, pp. 85–98.

[20] "The road to nanotechnology," April 2022, https://www.telecomreviewafrica.com/en/articles/features/2749-the-road-to-nanotechnology

[21] M. Hlophe and T. Hillie, "Chapter 40 – Challenges to implementing nanotechnology solutions to water issues in Africa," *Micro and Nano Technologies*, 2014, pp. 611–621.

[22] M. S. Mufamadi, "From lab to market: Strategies to nanotechnology commercialization in Africa," *MRS Bulletin; Warrendale*, vol. 44, no. 6, June 2019, pp. 421–422.

[23] K. Beumer, "Travelling risks: How did nanotechnology become a risk in India and South Africa?" *Journal of Risk Research*, vol. 21, no. 11, 2018, pp. 1362–1383.

[24] N. Z. Nyazema, J. T. Chanyandura, and P. O. Kumar, "Nanomedicine and regulatory science: The challenges in Africa," https://www.frontiersin.org/journals/biomaterials-science/articles/10.3389/fbiom.2023.1184662/full

[25] "Nanotech diplomacy: Opportunities and challenges for international cooperation," April 2024, https://unctad.org/news/nanotech-diplomacy-opportunities-and-challenges-international-cooperation

[26] C. F. Schutte and W. Focke, *Evaluation of Nanotechnology For Application in Water and Wastewater Treatment and Related Aspects in South Africa*. South Africa: Water Research Commission, 2007.

[27] Z. Mazibuko, *Beyond Imagination: The Ethics and Applications of Nanotechnology and Bio-Economics in South Africa*. African Books Collective, 2018.

[28] T. F. Barker et al., *Nanotechnology and The Poor: Opportunities and Risks For Developing Countries*. Springer Netherlands, 2011.

[29] Mistra, *Beyond Imagination: The Ethics and Applications of Nanotechnology and Bio-Economics in South Africa*. African Books Collective, 2018.

[30] M. N. O. Sadiku, S. M. Musa, T. J. Ashaolu, and J. O. Sadiku, *Applications of Nanotechnology*. Gotham Books, 2023.

Index

https://doi.org/10.1515/9783112211984-011